STRESS AT WORK
A Sociological Perspective

Chris L. Peterson

POLICY, POLITICS, HEALTH AND MEDICINE SERIES
Vicente Navarro, Series Editor

Baywood Publishing Company, Inc.
Amityville, New York

Library of Congress Catalog Number: 98-11799
ISBN: 0-89503-190-6 (Cloth)

Library of Congress Cataloging-in-Publication Data

Peterson, Chris L., 1949-
     Stress at work : a sociological perspective / Chris L. Peterson.
        p.   cm. - - (Policy, politics, health, and medicine series)
     Includes bibliographical references and index.
     ISBN 0-89503-190-6 (cloth)
     1. Job stress.   2. Industrial sociology.   3. Industrial hygiene.
     4. Job stress - - Australia - - Case studies.   I. Title.   II. Series.
HF5548.85.P483   1998
158.7 - - dc21
                                                                                    98-11799
                                                                                        CIP

Contents

iii

Acknowledgments

I am indebted to colleagues at La Trobe University, especially Professor Evan Willis and Professor Brian Graetz for their support of this work, and to my wife Jan.

I would like to acknowledge the Australian Council of Trade Unions (ACTU) campaign on stress which is heightening awareness about the growing problem of stress in Australian workplaces and producing positive changes. The ACTU's recent workplace survey has drawn attention to workers' stress resulting from management control and practices.

INTRODUCTION

The Social and Political Context
of Stress

The trend in most western capitalist societies in the late 20th-century has been to emphasize high productivity in a competitive economic environment. Due to the upsurge of economic rationalist thought and concurrent economic downturns, management has significantly increased its control over production processes. While many perspectives can shed light on the direction and nature of these trends toward increased managerial dominance and control, the Marxist approach is particularly clear in unraveling the forces leading to the forms of control now evident in many work organizations. That hegemony is alive and well as most union movements have lost some direct influence over the organization of work.

The struggle for dominance of the classes is still strong, though masked by a rhetoric that emphasizes management's need to exercise increased control to grapple with economic problems and a lack of control, albeit in the interests of all. Yet, still there is a division between those who own the means of production and those who sell their labor.

The economic and sociopolitical framework within which the broader process of production is carried out incurs certain social as well as personal costs. The organization of work and dominance by management maintain conditions at work that are conducive to occupational stress and ill-health. However, greatest ill-health among occupational groups is likely to be in the lower occupational and class groups (for example, blue-collar workers). They are most at risk as they work chiefly to fulfil goals that are based on management's interests and not necessarily their own. These interests are bound up with management maintaining its own control.

This book has a number of aims. Specifically it focuses on the causes of stress and ill-health in the workplace and differential distributions among occupation and gender groups. Level of management control is an important factor affecting workers' experience of work. There are a number of ways in which management exerts control over the workplace: this can affect many levels of workers' experience. The ability of workers to effect changes to management's control at work is crucial to understanding how employees can influence outcomes of stress and ill-health. In addition, this book addresses the factors at work that lead to

1

stress and ill-health among blue-collar workers compared with white-collar workers. The extent to which management control has an effect on differences in stress and ill-health between women and men is also examined.

STRESS:
SOCIOLOGICAL CONCERNS FOR INVESTIGATION

The problem examined in this book is investigated primarily from a sociological vantage point. It looks at whether differences exist in the experience of occupational stress and ill-health between blue-collar (working-class) workers and white-collar (more middle-class) workers. Management exerts control over the labor process, and this can be seen as a crucial factor in the degree to which important needs are not met by workers. Stress and other sociopsychological states such as low self-esteem and low personal mastery can result from experiencing insufficient influence. These states in turn can lead to ill-health at work. An important question addressed in the book is the extent to which worker ill-health in organizations in capitalist society can be seen as preconditioned by management control.

The symptoms of occupational stress have been evident for a long time. However, stress has only recently been placed on the agenda as a cause of serious concern for workers' health, productivity increase, and effective management in industry. The increased interest has come about partly due to an increase in the occurrence and effects of stress. Medical practitioners, government agencies, unions, and industry leaders have also become more aware of the incidence of occupational stress and the implications for people's health as well as productivity levels. Major impacts have resulted from changes in legislation on occupational health and safety and in compensation laws.

Burawoy (1), a sociologist, draws on the idea of a political economy and argues that a management-controlled labor process is embedded in the workings of the broader political economy of a capitalist society. Occupational stress needs to be seen as a by-product of that broader sociopolitical activity. Stress is also part of a wider problem of owner and manager control of the labor process. As Braverman (2) argues, from the labor process view, the organization of work can have negative outcomes for workers. Frederick Taylor's (3) scientific-management-controlled production processes, developed from the 1920s, have had major effects on modern work organization and have created a deep-seated problem as causes of occupational stress. The issue of control of the labor process by management needs to be questioned.

Studies of occupational stress typically follow two predictable paths. Most psychological studies implicitly support management's right to control. Many of these are attitudinal and attempt to identify the personality characteristics and attitudes of employees who suffer stress. Reductionist psychological studies present more extreme examples of focusing on individuals and their problems.

Many psychological studies focus attention on the benefits of stress management techniques as a way of reducing stress rather than on issues of conflict between management and workers. They also tend to ignore that work conditions are likely to make some occupational groups more prone to stress than others. Labor has fewer studies to support its orientation. These studies are mainly sociological and are more concerned with specific work stressors. They tend to focus on recommending job redesign, participative management, and more effective compensation schemes for stressed workers. Some of these studies deal specifically with fundamental conflicts between management and labor as the cause of work stress.

Since the early 1970s a sociological approach called the labor process approach has addressed issues of broader structural control as a cause of occupational stress. These studies emphasize labor's lack of control over the labor process as a principal cause of stress and consequent ill-health (see Coburn (4), Garfield (5), Navarro (6), and Schwalbe and Staples (7), among others). Based on the work of Braverman (2), labor process theory argues that the structural framework within which the process of production is carried out maintains conditions at work that are conducive to occupational stress.

Management's attempts to exert control over the labor process lead to alienating working conditions in which those with least influence fail to satisfy important needs or have insufficient resources to deal with excessive demands. These workers are likely to experience stress and consequent ill-health. However, many labor process writers identify a lack of influence on the job itself as a cause of stress and ill-health. Management in fact exerts a considerable control over most areas of workers' experience: these other outcomes of management's control may lead to greater stress. They include the effects of bureaucratic control over the structure of work and work relations, and the climate of attitudes and values in the organization, typically those of management toward workers.

When management control leads to alienating work and employees fail to satisfy important needs, stress is likely to result. In addition, when sociopsychological resources (self-esteem and mastery) are low, employees may not be able to deal effectively with undue demands or with unsatisfied needs. Ill-health is likely to result. Those in the lowest occupational positions, who have least influence over corporate decisions and policy, probably share little agreement with the goals of management. The experience of stress and ill-health is likely to be largest for these workers.

When management exerts substantial control at work, the ability of workers to influence or change the managerial imperatives upon which the capital-controlled labor process is based is crucial to occupational ill-health. The sociopolitical processes within which work is carried out—or as Burawoy (1) notes, the conditions of the political economy—establish the preconditions and structures for managerial control of the labor process. This control normally disadvantages lower level workers. A precondition of management control in capitalist society is a degree of worker ill-health (see Schatzkin (8)).

Occupational stress occurs as a consequence of the conflict of interests in industrial society: it is part of the risk imposed by the process of capital accumulation. At the broadest level stress results from management's control of the labor process. It results in particular from management establishing its prerogative to pursue its interests. It also arises from conflicts inherent in a work system that does not properly sustain the needs of workers.

Stress has relatively recently become recognized as a legitimate basis for claim as an occupational illness or injury in Australia. Under Workcover legislation in Australia, work compensation provisions allow for legitimate claims for stress-related illness and injury. In the past stress had largely been treated as one of a number of psychosomatic illnesses. Stress as well as conditions such as occupational overuse syndrome (formerly repetitive stress injury, RSI) still remain highly contentious issues in compensation courts in Australia. The conflict of interest between management, the corporate sector, and the claimant involves an interplay between medical practitioners and insurance companies supporting the corporation and the sometimes conflicting positions of labor. The underlying conflicts between the ideologies of capital and labor are played out in the assessment centers and compensation courts. In addition, bias in health care approaches by physicians and other practitioners can focus the problem of stress as an individual concern and redirect interest away from some of the basic causes of illness.

In this book I review physiological, psychological, and sociological approaches to understanding the causes of occupational stress. While physiological and psychological approaches offer important insights into the stress response and into people's reactions to negative experiences, sociological research offers important insights into understanding the context and causes of stress and ill-health at work.

The book presents results based on an examination of a series of relationships between variables in a causal model through a study of a U.S.-owned manufacturing organization in Australia. The causal model investigates the effects of ascribed and achieved statuses (age, gender, level of education, partnership status, and number of children), and the effects of occupational position and length of service in the one job, on stress and ill-health outcomes. In addition it tests the effects of these independent variables (ascribed and achieved status, and employment characteristics) on self-esteem and personal mastery. Finally, the degree to which stress, self-esteem, and mastery mediate outcomes of ill-health and minor psychiatric impairment is examined. Occupational and gender differences are also investigated. The model is tested to examine the degree of support for the existing literature and theory that explain stress and ill-health outcomes through individual experience and sociostructural factors.

The study examines the stress and ill-health experiences of 242 women and men through a questionnaire-based study. The study questionnaire was completed by respondents ranging from workers on the factory floor to members of the board of directors.

THE AIMS OF THE BOOK

The book has four major aims. First, it focuses on an area of stress that has been largely neglected in the growing field of stress research. Relatively few studies have examined the incidence of occupational stress in a broader sociopolitical structural context. Sociostructural and broader class relations explain important aspects of the causes, distribution, and consequences of occupational stress. Second, stress models are examined based largely on the pioneering work of Hans Selye (9), tracing the shift in emphasis from descriptive models based purely on physiological factors to the development of processual models. While physiological models provide an important understanding of the process and structure of the stress response, purely physiological models offer the least effective explanations of the causes of stress. Later psychological models help to explain some of the differences between individuals in the experience of stress. However, they neglect to explain the influences of the broader social and structural context within which occupational stress takes place. Later sociological models account for the different meanings that people give to stressful experiences and their effects on the physiological and psychological process. They also account for the influence of social structural factors in explaining the stress process.

The third aim is to investigate the role of a labor process approach in explaining the broader social context of work, stress, and ill-health relationship. The role of management control over the labor process is treated as an important contributing factor to stress, other sociopsychological states (e.g., self-esteem and mastery), and ill-health. The degree of influence and control that employees have over their jobs and work environment helps to explain the occupational stress and ill-health process.

Sociological writers in the labor process tradition have argued that management's control over the labor process leads to workers' experiencing alienating work. Those workers with least influence are likely to experience the greatest levels of alienation (see Coburn (4)). Alienating work experience is likely to lead either to important needs not being met in the workplace and on to workers not having sufficient resources to deal with excessive demands. Stress can lead to ill-health and an accompanying lack of sociopsychological resources (low self-esteem and low personal mastery). Therefore, important personal resources for dealing with the effects of unfulfilled needs or excessive demands are lacking. In addition, basic conflicts between the owners of capital (owners or managers) and labor have meant that the process through which stress has been recognized as a condition for concern and for compensation has been rather slow and difficult.

Fourth, the book describes the development and testing of a causal model linking negative work experience to outcomes of stress, low self-esteem, and low mastery, and ultimately to ill-health. There are few studies that link all of these concepts together. Still fewer studies link stress and ill-health to the broader social structure: these studies tend to be sociological. The labor process approach treats alienating work experience as being a result of little influence and control over the labor process. Workers in lower level jobs are likely to have higher levels of stress due to a lack of important needs satisfaction and to limited resources for dealing with demands at work. This largely results from insufficient opportunities to exert control. Similarly these workers are likely to experience poorer self-esteem and less personal mastery than those in higher occupational groups. This can be due to more alienating work together with less positive and less rewarding past social experiences. These sociopsychological resources are important in dealing effectively with negative work outcomes. Ill-health can result from demands and pressures and from a lack of effective resources that come from occupying lower positions in the work structure. The effects of sociostructural factors on stress and ill-health are examined through differences between occupational groups and between women and men.

OCCUPATIONAL STRESS: A DIVERGENT CONCEPT

A great deal has been written about occupational stress. Public knowledge and legislative interest have increased during the last decade in many western countries. It is still to some degree a disease with a contested terrain: that is, there is still not a wide degree of acceptance of the condition.

However, unless its causes are seen to lie in the structure of work relations and in broader class relations, occupational stress runs the danger of being treated as an individual phenomenon. In this book I argue that management control and the process of management that develops out of that control are a major cause of stress. This is an outgrowth of the Marxist idea of the domination of the working classes by the owners (and managers) of the means of production. That domination leads to poorer levels of work experience, greater stress, and ultimately worse levels of health for working-class workers. Labor process approaches which argue that increased management control over the labor process leads to increased alienation for those with fewest resources and least influence at work are a useful theoretical framework for understanding the causes of stress. However, that argument can be extended to include the effects of alienating work features throughout other aspects of work experience beyond the effects of task-related activities.

This book explores the relatively underdeveloped attempts at explaining the phenomenon of stress and ill-health through links with the sociopolitical structure of the broader society. It also explores the differences in stress and ill-health between working-class and middle-class workers, and differences between women's and men's experience at work. The structure of work relations and their sociopsychological and ill-health effects result from entrenched conflicts in class relations. The work by Harry Braverman (2) on the effects of the capitalist labor process is an important starting point for such an evaluation.

Many studies of the effects of work experience on stress and ill-health have looked at the effects of a lack of control over the task itself. First, an evaluation of the effects of management control focuses on a lack of influence and control at this level, together with its effects on other levels of work experience. Also examined are the effect on outcomes of management's influence over the structure of work and work relations, and the influence of managerial dominance on the climate of negative values and attitudes especially toward lower level employees. Each of these effects examines specific outcomes of class conflict in work experience.

Work on stress derives from a wide range of disciplines including psychology, sociology, biology, and medicine. Frankenhaeuser says that "research on human stress . . . is the meeting place for several disciplines" (10, p. 213). The sources of stress research and writing have evolved from widely differing areas of interest, and consequently the area of stress has a number of problems both in concept and definition. There are diverging views in the literature on the merits of using the concept of stress. For example, Hinkle (11) suggests that this concept is no longer adequate for understanding the relationship between people and stressors and consequent ill-health, nor is it even helpful: this is in contrast to other writers who emphasize the stress concept's inherent value. Lumsden, for instance, argues that the stress concept (12, p. 191):

> is one of the most significant and integrative concepts ever developed in the social and biomedical sciences, and that its potential as a prime intellectual tool for not only understanding, but also explaining individual and collective human behaviour and disorders has not yet been fully realised.

Cox and Mackay (13) argue that while the concept of stress is useful in that it brings together literature and research on many different issues and problems, the conceptualization and use of the term have differed widely. Stress research has focused on significantly different aspects and has approached problems in a variety of ways. The concept of stress has been used to identify a dynamic psychosocial process intervening between stimulus and response. Mason has argued that the definition and conceptualization of stress have been confusing (14, p. 9):

> The disenchantment felt by many scientists with the stress field is certainly understandable when . . . the term stress has been used variously to refer to "stimulus" by some workers, "response" by other workers, "interaction" by others, and more comprehensive combinations of the above factors by still other workers.

Selye (15) uses the term stress to mean a state within the organism; others refer to stress as an event or stimulus tapping the coping resources of the individual. These various definitions can be a source of constant debate, but need not be overly problematic if stress is seen as a term for the whole area of study. Lazarus's contention that the term stress is not a "stimulus, response or intervening variable, but rather a collective term for an area of study" (16, p. 27) seems appropriate. It is important, however, to clarify an operational definition, and rather than attempting to add one more definition of the term, I will adopt here conceptions previously developed and used. Levi suggests that stress "refers to a process in the body, to the body's general plan for adapting to all the influences, changes, demands and strains to which it might be exposed" (17, p. 1), both physical, mental, and social. Based on Levi's suggestion of a process of adapting to outside influences, stress can be seen as a state of discomfort or ill-ease, an unpleasant state within the individual created by the individuals' inability to satisfy needs and desires in relation to demands made and resources available in the environment.

Stress, in this book, is treated as depending on individual perception; a state in which, in their course of action, individuals are unable to satisfy important needs in their interaction with the demands and available resources of their environment. The roles of other sociopsychological states—low self-esteem and low mastery or sense of control over external events, among other factors—are important in stress outcomes. Self-esteem and mastery, the evaluative parts of the self concept, have interactive effects with stress in which they facilitate the process of adapting to stressful experiences and can mitigate or contribute to their effects on ill-health.

This work initially was based on a strong interest in democratization of the workplace as a way of approaching some of the problems that have their roots in structural and therefore class relations. Democratization of the workplace in countries such as Sweden and Norway has shown that significant improvements in the quality of work life can be gained, together with increased benefits for stress and ill-health. However, with cost-containment policies becoming dominant, increased integration of organizational activities, and emphasis on the responsibility for individual workplaces to achieve solutions given the economic downturns during the decade, we must seek new solutions for effective organization and reduced health impairment from work.

ORGANIZATION OF THE BOOK

Chapter 1 deals with developments in stress models since the pioneering work of Hans Selye. These developments have seen some close integration of psychological explanations of stress with purely physiological approaches. While purely physiological models of stress show a clear relationship between noxious external situations, stress, and ill-health, they are far from explaining the full nature of the causes of stress. Psychological models have shown that stress occurs differently among individuals and depends upon a cognitive and emotional response as well as a physiological response. However, psychological approaches do not address any of the sociostructural context within which stress takes place.

In Chapter 2 a review of the literature on occupational stress examines the role of a lack of control and other negative work experience. Much of this research, however, is psychological: while identifying noxious work factors as important, it fails to examine the effects of broader sociostructural factors on work stress and ill-health. Chapter 3 examines the contributions of sociological explanations to understanding the stress process. Sociological approaches identify those structural and social factors that help to explain the occurrence of stress among different groups in society. In addition, the labor process approach to explaining stress at work is introduced and developed.

Chapter 4 reviews later developments of labor process approaches, which show the role of new technological development in enhancing management's control. Management's increased control through the development of strategic plans and policy creation is also outlined. The concept of alienation is shown to be important: it places the experience of stress in a comprehensive structural framework. This helps to demonstrate more clearly how management's control over the labor process leads to stress and ill-health. Chapter 5 is a review of research based on labor process approaches, identifying the causes and consequences of stress and ill-health for a number of different occupational groups. I also discuss the broader sociopolitical context of stress and ill-health. The various roles played by the state are outlined: the state's role in supporting management's right to exert control at work; the negative effects of the state's control on workers' jobs; and the role of the state as an employer and in affecting changes in the labor process.

Chapter 6 outlines the methodology used in the study of stress and ill-health outcomes. The causal model of stress and ill-health discussed here forms the basis of the analysis of results in the following chapters. I present in this chapter all the major dependent variables used in the study, clarifying the major measures used. Chapter 7 outlines the results of tests of the major effects of variables in the causal model on alienating work. The major determinants of alienating work experience are established through the use of regression analysis. Occupational and gender experiences are compared to test for the effects of management control.

In Chapter 8 I examine the major factors influencing stress. The first major hypothesis is that "as occupation level decreases, stress is likely to increase." Sociostructural influences are examined through a comparison of occupational and gender groups. The chapter presents a causal model testing the effects on occupational stress of ascribed and achieved statuses, occupational statuses, and negative work.

Chapter 9 examines the effects of factors in the causal model on self-esteem and personal mastery. A second major hypothesis tested is that "as occupational position decreases, self-esteem decreases." In Chapter 10 the effects of sociostructural variables are tested on somatic and emotional symptoms. The final stage of the causal model in the study is also examined. That is, all previous factors (ascribed and achieved statuses, occupational statuses, negative work, and stress and the other sociological states) are tested for their causal effects on frequency of somatic and emotional symptoms. A third major hypothesis tested is that "as occupational level decreases, ill-health symptoms will increase." The fourth hypothesis tested is that "as stress increases, so will frequency of ill-health symptoms."

In Chapter 11 I investigate the factors influencing minor psychiatric impairment (Goldberg's General Health Questionnaire, GHQ). As in the previous chapters that present results, sociostructural effects are determined through comparing occupation and gender differences. In addition, I compare psychiatric impairment scores with changes in stress, sociopsychological variables, and job satisfaction. The concluding chapter draws together research and theory and the results obtained in the study. I present a number of recommendations for developing research programs and outline implications for health policy and industrial health practice.

CHAPTER 1

Stress as a
Psychophysiological Process

In this chapter I trace developments in the concept of stress, as a physiological reaction and as a psychophysiological process. This historical review covers presociological inquiries of stress. It examines many of the debates on the meaning of stress, based on early differences in the meaning of the stress response. While these approaches have a number of limitations, an examination shows how early differences in biophysiological and later psychological research contributed an important understanding to the stress response.

While earlier, purely physiological approaches provide information about bodily reactions to stress, they do not provide a sufficient understanding of a broad range of processes involved in the stress response. Parallel and more recent psychological research provides some important insights into why different people experience differing amounts of stress from the same stressors. These findings make important contributions to our understanding of the psychophysiological nature of stress.

However, psychophysiological approaches fail to account for a broader range of sociocultural factors, a topic addressed in Chapter 2. Yet these psychophysiological processes do provide an important basis for our understanding of stress, albeit at the individual level of personal experience.

While the early biophysiological and psychological researchers have made important contributions, our knowledge about the stress response has recently changed in some important ways. Technological developments in testing procedures and equipment have also produced major changes in our understanding of this response. Yet certain limitations to a thorough understanding of biophysiological and psychological processes remain. We need to employ a critical framework when dealing with various assumptions about the biophysiological aspects of the stress response that require questioning.

The chapter covers three issues. The first is a review of the pioneering work of Hans Selye and a critical evaluation of early developments in the stress concept. I then review the influence of psychologically oriented research into the effects of emotional response on physiological reactions and evaluate it for its influence on the stress concept. Finally, I discuss the physiology of the stress response, based on the above approaches.

THE STRESS RESPONSE:
THE EARLY INTEREST IN PHYSIOLOGICAL MODELS

Hans Selye was one of the most notable contributors to our understanding of the stress response. He was born in 1907 in Vienna; he gained his medical degree in 1929 from the German University of Prague and later his Ph.D. from the same university. He spent the later part of his life in Canada. In 1935 his work on what he thought to be the discovery of a new sex hormone led him to "discover" the stress syndrome. His first publication on stress, in *Nature* (9), sparked a new and considerable interest in the field of stress research. Since then he has published extensively. He was recently the president of the International Institute of Stress at the University of Montreal in Canada.

Even though the term stress had been used prior to Selye, notably by Cannon as early as 1914, it is mainly Selye who has been associated both with the initial development of links between socially produced states of tension and ill-health and with the increasing acceptance of the concept of stress.

While Selye initially used the term stress to denote external agents acting on the organism, he developed the notion of stress occurring within the organism. He maintained that stress is a physiological response: "the state manifested by a specific syndrome which consists of all the nonspecifically induced changes within a biologic system" (15, p. 54). These changes, he argued, are caused by damage or function. Selye developed and clarified the concept of physiological reaction to the demands made on the organism, and he developed the concept of the general adaptation syndrome (GAS), a reaction produced by a variety of unpleasant agents. The alarm reaction is the initial physiological response to a stress-producing event. In the resistance stage, adaptation to the stressor takes place. Physiological reactions change, and the body's resistance is higher than normal. Eventually, after long-term exposure, the body's capacity for adjustment declines and exhaustion occurs. At this stage, Selye says, the individual is prone to develop illness or may even die. After many years of experimentation Selye concluded that the "body's adaptability . . . is finite (15, p. 38).

Two physiological pathways are activated under stress. One, the "hypothalamus-autonomic nervous system-adrenal medulla" results in adrenalin and noradrenaline secretion. Adrenaline is an important hormonal response in the alarm reaction. With an initial stress or shock, messages are sent through the sympathetic nervous system to increase alertness and physical defenses. Blood pressure and heartbeat increase. Otto (18) refers to this as an increased propensity for fight or flight. The stimulation of adrenaline and noradrenaline secretion increases the level of energy in order to deal with increased demands.

Of equal importance if stress is maintained is the "hypothalamus-pituitary-adrenocortical axis which maintains a state of bodily equilibrium. According to Selye, this also produces many disease manifestations. The pituitary gland regulates adrenocortical activity: when the hypothalamus (which is connected to the pituitary) is stimulated, ACTH (adrenocorticotropic hormone) is secreted into the blood, which stimulates secretion of corticoids from the adrenals; this results in sugar for energy. Selye suggests that while "every disease causes a certain amount of stress since it imposes demands for adaptation upon the organism . . . stress plays some role in the development of every disease" (15, pp. 46–47).

The principal physiological relationships recognized through the work of Cannon (19, 20) and Selye show that when the bodily system becomes aroused, the autonomic nervous system is activated. When both the sympathetic and parasympathetic divisions are activated, the appropriate adjustments occur to the body's functioning. The results are deeper respiration, increased heart rate, dilation of the pupils, and constriction of blood vessels. The next step is increased adrenaline production, which raises blood pressure, produces glucose for extra energy, and releases fatty acids for glucose production. With prolonged stress, the adrenal cortex (the outer part of the adrenals) produces hormones such as cortisol and other glucocorticoids which reduce inflammation and pain. If stress is prolonged, however, the benefits of cortisol decrease and the organism develops a greater susceptibility to illness.

PROBLEMS WITH THE INITIAL CONTRIBUTIONS

There are several problems with these initial conceptions of stress, however, despite their importance. A major problem is that many researchers have now challenged Selye's concept of the nonspecificity of the stress response. In his definition of stress Selye refers to nonspecificity: stress, he argues that is a "state manifested by a syndrome which consists of all non-specifically induced changes in a biological system" (15, p. 54).

The problem of why people react differently to stressful situations or stimuli, and why certain diseases develop in some individuals and different or no diseases develop in others, is not accounted for. Selye's view was that given sufficient intensity, different stimuli are going to produce the same response pattern of stress. He argued that "contrary to previously held opinion, stress is not identical to emotional arousal or nervous tension" (21, p. 15). He further argued, "from the point of view of its stressor activity, it is even immaterial whether the agent or situation being faced is pleasant or unpleasant; all that counts is the intensity of the demand for readjustment or adaptation that it creates" (21, p. 14). Selye did respond to critics, however, by saying that the stress syndrome is nonspecific by definition: however, depending upon conditioning which can influence the reactions of organs, stressors may have different impacts on different people.

Selye has attempted to deal with the problem of specificity in a number of ways. First he argues that the effects of stress can be good or bad, depending on eustress and distress, and those based on negative experiences are much more likely to cause diseases. Eustress is stress producing pleasant effects, and distress produces unpleasant reactions. This is despite the fact that stress in Selye's general adaptation syndrome produces limited, stereotypical physiological responses. Pollock (22), in her review of Selye's contributions, provides a number of useful criticisms of his assumptions. She argues that there is a shift from a one-dimensional approach to a thinking, cognitive approach. Pollock rightly points out that this involves an "incongruous and infathomable shift . . . when having to account for how an identical set of physiological responses can have radically different consequences for health" (22, p. 385), and this involves the need for perception, evaluation, and response. It is difficult to see how Selye can maintain the need for a concept of nonspecificity while acknowledging the role of psychological factors.

In Selye's second attempt to deal with the problem of specificity he argues for a distinction between specific and nonspecific effects by showing that different stressors will produce specific effects on organisms as well as nonspecific, general adaptation syndrome effects. This distinction really only confuses and complicates a difficult process, particularly when involving psychological factors in the stress process.

Seggie and Brown argue that "accumulated evidence suggests and hormones manifest differential sensitivity and responsivity suggesting that the response mechanisms are pathway specific" (23, p. 277). They argue that the adrenal cortex may be the most generally responsive pathway and may respond in the least specific way. There are times when other hormone-secreting glands and not the adrenal cortex respond.

COGNITIVE AND PSYCHOLOGICAL ELEMENTS
IN THE STRESS PROCESS

A number of researchers have argued the need for including cognitive and psychological elements in the stress response. Many more recent studies have shown the effects of stress on psychological factors (e.g., Ferrandez and coworkers (24), Heuther and coworkers (25)). Researchers such as Mason (26), Monet and Lazarus (27), and Cassell (28) have argued that nonspecificity in the production of illness has been overemphasized in Selye's work. Lazarus suggests that environmental demand, the quality of the emotional response, and the coping style mobilized will influence the nature of the stress disorder. Mason argues that there are great difficulties in trying to separate effects of physical stimuli from psychological reactions to them, and that regarding adrenocortical activity in particular, the stress concept should be seen as behavioral rather than physio-logical. Mason argues further that when psychological reactions to physical stimuli are immunized, "it now appears that the pituitary-adrenal cortical system is not stimulated in non-specific fashion by these [physical] stimuli which are generally regarded as . . . appreciably disturbing to homeostatic equilibrium" (26, p. 24). Levi (29) mentions that the mechanisms influenced by psychosocial stimuli are mental processes; endocrine, lymphatic, and immunoreactive processes; and other physiological processes. Future research may show a much higher degree of central nervous system functioning with psychological stimulation than has yet been realized.

Monet and Lazarus argue that early research "minimised the ethological significance of psychological stress factors and instead, emphasises the role of stereotypic bodily reactions . . . to any tissue assault in increasing susceptibility to all illnesses," adding that before hormones adjust, the subject must appraise the situation as threatening and that "these adjustments are specific to specific threats" (27, p. 7). Egger (30) also argues that evaluation of stress-inducing situations is important, especially in conditioning reaction to stress. Levi suggests that psychological reactions produce a set of biochemical reactions which may or may not influence physiological mechanisms: these consequently may have an influence on either the suppression or the production of disease. He refers to his earlier research as evidence for the range of effects of psychosocial influences on neuroendocrine function. Sympathoadrenomedullary, adrenocortical, and thyroid activity can all be influenced by psychosocial stimuli. Levi concludes that "a very large number of physiological processes are influenced, directly or indirectly. . . . we know that psycho-social stimuli cause physiological changes, which in turn could lead to precursors and disease" (29, pp. 19–20). Christiansen and Jensen (31) suggest that inadequate responses to psychosocial stress and inappropriate coping approaches may be even more harmful than large secretions of stress hormones (such as cortisol). However, these hormonal responses may occur as a result of inadequate psychological adjustment.

Other research has shown the effects of physical stimulation on psychological outcomes, such as studies conducted by Melamed and Bruhis (32) on the effects of industrial noise on irritability. Some of this research is examined in Chapter 2.

Certain other problems exist, based on the pioneering work of Selye, Wolff (33), Hinkle and coworkers (34), and others. First, the connection between stress and disease is well known, but understanding the strength and direction of that connection needs further research. Second, the importance of perceptions in shaping the stress reaction and response is understood, but this also requires further research—although there is increasing agreement on the role that it plays. Third, to date, models of the stress response have omitted potentially important hormonal responses, and we are still a long way from developing a total model of the stress reaction. Finally, not only did purely physiological research neglect the role of individual personal variables in explaining the stress response, physiological models largely ignored the impact of social, cultural, and economic variables.

These models, therefore, have provided only a partial picture of the stress response. I have shown that Selye's concept of nonspecificity ignores cognitive and perceptual processes that have been found in psychological research. In this sense his account of the stress response is limited: it neglects to account for different people responding in different ways to the same stressor. Also there are important reasons for considering stress to be a psychophysiological process, not just a purely physiological process. Research by Mason (35–37) and Levi has shown that subtle hormonal changes occur as a result of cognitive and emotional responses. Ferrandez and coworkers (24) have associated psychological stress with greater concentration of stress-related hormones.

Christiansen and Jensen (31) have examined the effects of psychosocial stress on plasma neopinephrine, while Groenink and coworkers (38) have studied the role of mediating agents in emotionally induced stress hormone reactions. However, the use of both physiological and psychological approaches to the stress response needs strengthening, based on the results of some very strong evidence.

TWO APPROACHES TO RESEARCHING STRESS

Developments in the conceptualization of stress and directions of stress research have occurred, with two strands of research developing side by side: research that emphasizes the physiological bases of stress and research that emphasizes the role of psychological factors.

The history of interest in the stress concept and stress research lies soundly within the domain of physiology, despite the fact that some early interest was shown in emotional response as a basis for stress. Mason argues that the late 1950s saw the beginning of a decline in interest in stress in the physiological field, and that decline has continued. At the same time there has been an overwhelming interest in psychophysiological approaches which have added new research techniques and concepts to the earlier, predominantly physiological explanations of the stress process.

The concept of stress was used as early as 50 years ago by Cannon and Selye. Developed by Selye, the concept of a general adaptation syndrome, with virtually any noxious agent or stressor producing the same or similar responses, has persisted relatively unchanged over this period. Despite the wide range of contradictory propositions about the stress process, as described earlier, most researchers and theorists still accept these as basic tenets. Mason (37) and later Burchfield (39) and others (4) have demonstrated the need to consider emotional responses beyond purely physiological stimuli in the conception of stress.

Psychological and physiological factors have converged in stress research over the past two decades, and there is general agreement that the stress response cannot be adequately explained without accounting for their dual role. Mason observed that emotional influences, particularly as involved in the regulation of corticosteroid levels, were almost universally underestimated by early stress researchers, including Selye. Cannon, however, showed that the sympathetic-adrenal medulla system was influenced by psychological factors, but it was a long time before this found credibility in mainstream stress research.

Mason (36) describes Selye's early experiments as examples of the role played by psychological reactions, but is critical of Selye's lack of attention to the influence of emotional and cognitive stimuli on endocrine activity. In fact, when Selye describes the response of the pituitary-adrenocortical system to different stimuli, Mason argues that it involved some emotional reaction. When Selye refers to "the usual response of the organism to stimuli such as temperature changes, drugs, muscular, exercise etc" (9, p. 4), these may all involve a degree of emotional or psychological reaction.

Selye referred to the influence of emotional stress on the alarm reaction, but certainly downplayed its significance. Mason (35) argues that the possibility of emotional reaction was raised quite independently by Selye in his work on the pituitary-adrenocortical response. Physiologists downplayed the role of psycho-logical factors. Even though the obvious separation between the approaches to physiological and emotional factors has been largely bridged, Mason's arguments have been echoed by Vingerhoets and Marcelissen. They argue that "interesting hypotheses can be formulated concerning the relationship of the catecholamine and corticosteroids and psychological concepts such as active and passive coping, but until now these substances have hardly ever been investigated together (40, p. 281). More recently, Huether and coworkers (25), Groenink and coworkers (38), and others have looked at the relationships between a range of stress hormones and emotional and psychological outcomes.

Henry and Stephens (41) show that if there is a positive appraisal of a stress situation, the amygdala and the sympathetic-adrenal medullary system will come into play. If, however, the appraisal is of a threat that is too great, the hippocampal-pituitary-adrenocortical system is activated, resulting in a subor-dination response. Weiss (42) has demonstrated the importance of emotional factors. He performed his experiments on rats, showing that the number of times that coping modes were employed correlated directly with the amount of tissue damage. Weiss and Baum and coworkers (43) are among the few researchers who have shown the effects of emotional reactions on biological processes in the stress reaction, while Cahill and McGaugh (44) found a relationship between the emo-tional response based on stress and memory storage.

The biological processes activated in situations or stimuli perceived as manage-able by the person are different from those activated by the perception of an unmanageable situation (see Weinstein and coworkers (45), Bandura and coworkers (46)). Bush (47), who as early as 1962 had said that psychological factors contribute to physical stress, showed that the secretion and metabolism of cortisol were not significantly influenced by physical stressors unless they were accompanied by an emotional response. Vingerhoets and Marcelissen (40) also have reported a relationship between emotion and physiological response, in that men who prefer problem-oriented coping over emotional coping have higher plasma catecholamine and testosterone levels. Also, these authors argue that a loss of control brings into play the parasympathetic nervous system.

By the early 1970s Mason had concluded that we could now say that adrenocortical activity reflected subtle psychological stimuli. Cannon's (20) work also showed that the sympathetic-adrenal medullary system was influenced by psychological factors. Selye had observed that psychological stimuli can elicit a substantial pituitary-adrenocortical response, although he did not emphasize this in his discussions of the overall stress response.

There are some reasons why physiological and psychological approaches have remained distinct. First, research into the stress response has not been very interdisciplinary, even though our understanding of the response involves knowledge derived from many different approaches. Second, stress research arose from two quite different physiological bases. One focused on the action of the sympathetic nervous system, which in its measurement is less responsive to the impact of emotional and cognitive factors. The other focused on the workings of the hormonal system through the adrenal medulla. This latter stream of research took much longer to capture the imagination and interests of researchers. The hormonal system is more responsive to measures of emotional response.

Research has now shown a strong interrelationship between emotional factors and the physiological response. This is largely due to advances in technologies that can measure the hormonal effects of emotional and cognitive responses. Mason argues that "the question of how the many hormones in the body act together as a group to accomplish the homeostatic regulation of physiological processes is one of the major unsolved problems in endocrine physiology" (48, p. 791).

CONTRIBUTIONS OF STRESS RESEARCH FOCUSING ON EMOTIONAL RESPONSES

During the 1960s and 1970s much parallel stress research was being carried out in a number of different disciplines, but much parallel and complementary research in other disciplines was ignored. Mason, a critic of physiological stress research, favored a more holistic approach. He described the problem in the early 1970s, when he said of endocrine research that it (49, p. 325):

> is generally confined to endocrinological and physiological laboratories quite far removed from the scene of psychoendocrine studies proceeding in Departments of Psychiatry and Psychology. The separation of the two research areas is almost complete, even to the point of extremely limited communication of published findings between the fields.

He further argued that "the slowness with which psychoendocrine data were incorporated into physiological thinking stems largely from the traditional separation of the behavioural sciences from the rest of biology and medicine" (49, p. 325).

Advances were made more recently through the integration of research findings. When sufficient links are made between different fields, particularly those that deal with the activation of responses leading to endocrine activity and the paths used in these processes, the complexity of the stress response will be better understood. As early as 1971 Mason suggested that a consideration of distinct stimuli with connections between neuroendocrine systems and their receptors and with neuroendocrine pathways was required. He further argued that if this were possible, there would be accurate historical and scientific bases both for stress theory and for the work of Selye. However, there were other reasons for the slowness in incorporating emotional responses into the stress response. More recently, researchers have been able to show the presence of greater hormonal activity in the stress response (40, 50–55). A principal reason has been improvements over the methods used for conducting experiments before the early 1950s. Other important developments have occurred. Some studies have identified an increased susceptibility to infectious neoplastic diseases and allergies from the stress reaction (56).

Initial research into the stress response focused on the sympathetic-adrenal system as a defense against stress. Selye, followed by others including Wolff, introduced the concept of stress as a response to noxious agents. Cannon, however, looked at stress as a challenge to homeostasis of psychological and physiological processes and focused on the hormonal system and adrenal medulla. McCarthy and coworkers (56) addressed in some detail the issue of the two fundamentally different early approaches to stress. Selye's conceptualization of stress has been dominant for nearly 50 years. Apart from the problems of nonspecificity, much of Selye's work and the work of other early researchers was limited to looking at the adrenocorticosteroids as playing a major role in the stress process.

While Cannon showed the impact of the sympathetic division of the autonomic nervous system, a little known French scientist, Reilly, observed the role of the autonomic nervous system, which seemed to precede the activation of the neuroendocrine system. Selye's work, in focusing on adrenocortical substances, neglected some important elements of the stress response. Vingerhoets and Marcelissen (40) outline several criticisms. Mason argues that in the early 1970s, due to the work of Selye, "attention in this field was concentrated heavily upon the pituitary-adrenal-cortical system" (35, p. 566). He suggests that interest in Selye's concept led to a narrowing of focus in research on endocrine regulation, particularly in Cannon's work on the adrenal medulla. Other hormones are involved in the stress response.

Evidence supports a variety of stressful stimuli that produce a range of hormonal responses. The pituitary-adrenal axis is the best known. However, prolactin, growth hormone (57), and beta-endorphin (58), as well as vasopressin have been found to be influenced by stress. Ducharme and coworkers (51) also describe the effects of "chronic psychogenic or somatic stress on testicular function." Pedersen and coworkers (59) have looked at the roles of growth hormone and beta-endorphin on immunomodulation through exercise. Less is known about the thyroid and pineal glands (60) in the stress response, although they have been identified as being affected by external stimuli. Far less research has been undertaken on the response of these various glands, and Seggie and Brown (23) argue that less is understood about the details of their possible roles. How many hormones are involved in body regulation is not yet fully understood. Early research on stress was based on the assumption that one system was dominant in eliciting physiological responses; this assumption was due both to conceptual limitations regarding the physiological response and to the relatively unsophisticated research techniques.

Mason has argued in an extensive discussion and review of psychoendocrine research that endocrine organization involves all hormones: "perhaps broad, coordinated patterns of hormonal responses are associated not only with psychological stimuli but with humeral or physical stimuli as well (48, p. 791). He suggests further that "the view that endocrine systems are generally organised as a group of specialised largely independent agents, among which the responsibility for regulating different body processes is sharply divided, does not appear to be tenable at present" (48, p. 794).

LIMITATIONS IN RESEARCH TECHNOLOGIES

Certain technological developments have allowed for a more extensive exploration of the stress response. Catecholamines and corticosteroids can be assayed by new technologies not available before the 1970s. Vingerhoets and Marcelissen identify "other pituitary hormones like β-endorphin, M.S.H. and substances that may play a role" (40, p. 281) between the brain and immune system (see 59). Sophisticated methods have been developed for continuous blood sampling, for recording psychophysiological factors, and for measuring cardiovascular reaction. Perhaps the most significant changes more recently are techniques that allow the measurement of variables in the stress response under natural conditions, so eliminating the heavy reliance on laboratory research which always has difficulties, particularly when trying to describe human reactions based on results from experiments with animals. Over the past two decades inroads have been made into understanding the effects of emotional stimuli, partly because research technologies and instruments for measuring physiological outcomes have reached far greater levels of sophistication.

Kopin and coworkers (61) argue that research in the 1940s led to a concentration mainly on the hypothalamic-pituitary-adrenocortical system. However, catecholamine research and stress research continued quite independent of each other. As new methods developed to assay the catecholamines, the activity of the central nervous system became more clearly defined. This change (around the early 1960s) brought stress and catecholamine research closer together. Psychobiological researchers such as Frankenhaeuser (10) demonstrate this trend, as part of a more systematic and holistic approach to stress research (62).

Techniques used to assay specific and direct effects had not reached a level of sophistication in early stress research. Selye had developed his conception of the stress response prior to the early 1950s, before chromatography and isotope techniques (which made possible much more precise measurement of the effects of stressors) were developed. Cannon showed the influence of psychological stimuli on the pituitary-adrenocortical system, and Selye hinted at it. Mason (49) argues that improvements in the sensitivity of measurement devices since the early 1950s helped in identifying specific hormonal responses. New links in stress theory became possible, particularly the role given to emotional stimuli. Despite their conceptual limitations, however, much is owed to the earlier researchers for developing a comprehensive stress theory with a meaningful physiological base. Purely physiological approaches developed before either research measurement tools or detailed emotional and psychologically oriented research had been developed. As a result, stress models suffered from limited scope and content. The role of individual perceptual and appraisal devices was seriously neglected.

A great deal of research has been conducted on relationships between stress and health, and more recent models have established stronger causal links. These have led to more holistic approaches. However, for some diseases the links between physiological factors and disease are not clear—for example, for cancer. Mechanic argues that stress is a complex process and serious methodological difficulties exist in building a sound bridge between concepts of biological, psychological and social stress" (63, p. 297). Large methodological differences exist between the physical, behavioral, and social sciences. The methods of laboratory experimentation used in physiological and some behavioral research have led to a slow and uneasy integration of psychological, biological, and physiological research.

The narrow individualistic focus of much psychological research is incompatible with social sciences' study of individual behavior and action as part of a broader set of social, cultural, and political processes. It is difficult to achieve integration when research aims at short-term stress response or individual stress experience. Mechanic suggests that stress research has been far too concerned with establishing short-term cause-effect relationships between stressors and outcomes. Stress, he argues, is "frequently seen as a short-term single stimulus rather than as a complete set of changing conditions that have a history and a future"

(63, p. 297). Physiology deals mainly with a short-term stimulus-response relationship (even though the process of stress may occur over a period of time). Psychology has expanded the dimension of research; it is mainly concerned with short-term *individual* and personal responses to stimuli. However, sociology, which will be discussed in Chapter 3, has moved closest to treating stress in a context of social, cultural, economic, and political processes. Sociology focuses on longer term processes in which a number of forces act to produce a biophysiological response.

THE STRESS RESPONSE

Despite differences in research orientation, there is general agreement on the type of physiological reaction in the stress response. Levi (17) outlines the physiology of the stress process. The autonomic nervous system and the endocrine glands are the most important defenses of the body. He argues, following Selye's work, that the most immediate response in adaptation to stress is through the sympathetic and parasympathetic divisions of the autonomic nervous system. This results in the body's functional adjustment. Levi describes the next stage as (17, p. 7):

> an increased production of the stress hormone adrenalin from the adrenal medulla in response to signals from the sympathetic nervous system. This stepped up adrenalin production, together with signals from the hypothalamus (the anterior part of the brain stem) stimulates the pituitary to increase its hormone production. The pituitary hormones regulate the production of hormones by other endocrine glands, and these hormones in turn participate in various ways in the body's defence and adaptation reactions.

According to Levi, ACTH is an essential hormone in this process: it controls a number of other hormones, the most important being cortisol.

Selye identified an important problem in understanding the physiological response in stress. He found that no one agent could be called the first mediator of the stress response; he discounted adrenalin, noradrenaline, acetylcholine, and histamine. He argued that the first mediator may be an emotional reaction to unpleasant stimuli, thus triggering a physiological reaction.

The physiological reaction involves a stressor, activated through two pathways, one through the sympathetic nervous system to the adrenal medulla and the other a hormonal response. This occurs through the pituitary gland, and the secreted ACTH activates the adrenal cortex, as described below. While the adrenal medulla and cortex belong to the same gland, their reactions are entirely separate. Through the hypothalamus, the autonomic nervous system, the part of the nervous system that operates without conscious control, is activated. Thisleads to stimulation of the adrenal medulla, producing secretions of adrenalin and noradrenaline. The initial reaction of the autonomic nervous system is the stimulation of the nerve endings, producing an immediate response; this is later followed by hormone secretion by the adrenal medulla. The result is an increased heart rate, dilation of the pupils, increased blood pressure, and increased perspiration. An increased awareness in response to the stressor results in restricted blood flow to parts of the body other than those being affected.

The adrenal cortex, as noted above, is activated through the pituitary gland and ACTH secretion. This is a hormonal reaction, producing a group of adrenocortical hormones known as the glucocorticoids, the most important of which for the stress response is cortisol. Cortisol influences energy production, acid-base balance, growth, wound healing, and resistance to infection. It also stimulates production of carbohydrates and has effects on lipid, protein, sodium, potassium, phosphate, chloride, calcium, magnesium, and water metabolism. Cortisol produces an immune reaction inhibiting the inflammatory response. Other hormones participating in the stress response include growth hormone, prolactin, beta-endorphin, vasopressin, and testosterone, but less is known about their influence or degree of effect.

The main hormones of the stress response are adrenaline and noradrenaline (catecholamines) and cortisol (a corticosteroid). They mobilize the body's resources in the form of glucose and the release of fatty acids and amino acids. After some time, these bodily defenses can produce a general reduction in the organism's immunity.

In the general adaptation syndrome described by Selye, the relationships between the three main responses (adrenaline, noradrenaline, and cortisol) have varying effects. Although the corticosteroids and catecholamines are quite distinct, the effects of adrenaline, noradrenaline, and cortisol are important in the body's general response to stress and in attempts to nullify harmful effects. In the alarm phase, adrenaline, noradrenaline, and cortisol may be secreted. If stressful conditions continue, a complex process takes place. As Frankenhaeuser (10) has pointed out, psychological or cognitive factors come into play and influence hormone levels. Cortisol is secreted for different reasons than adrenaline and noradrenaline. Adrenaline and cortisol secretion result from different response patterns.

Selye also argued that early research concentrated on glucocorticoids whose role is to lessen harmful reactions to injury, and that later research showed two groups of steroids to play a role "even more potent in maintaining homeostasis" (64, p. 264). These are the catatoxic steroids, which maintain homeostasis by destroying pathogens, and syntoxic steroids which are tissue tranquilizers. Syntoxic steroids induce tolerance "which permits a kind of symbiosis with the pathogen" (64, p. 264).

While the physiology of the stress response outlined above is generally accepted, the role of emotional or psychological reaction in precipitating and maintaining the stress response is clearly important. There is sufficient evidence to show that continued high adrenaline and cortisol levels can produce tissue damage and eventually lessen the body's resistance and immunity to disease (41, 61).

Selye's work on stress has not been conclusively confirmed or refuted by researchers during the past decades. However, it forms an important basis for understanding stress and identifying important assumptions underlying the stress concept.

CONCLUSION

Selye, often considered the "father" of stress research, has certain unresolved problems in his presentation of the stress response. His quite mechanistic conception of a nonspecific stress process leaves one problem in particular unanswered: given the same stressor or stressful situation, why do people react in different ways? Some become ill and may die, while others remain quite healthy. Even though Selye acknowledges the role of psychological factors, he does not account for differences in responses between people. Neither Selye nor other proponents of purely physiological approaches adequately answer this question. A purely physiological approach ignores many of the differences in stress reaction, whether due to environment, type of response, or other factors. A more processual and holistic approach to understanding the stress reaction and response is needed, one that accounts for perceptual differences and differing evaluations of the stressfulness of events.

Selye's model of stress and other purely physiological models are furthest from accounting for the role of important variables such as socioeconomic status and social class. They describe stress as a personal, individual process. Later psychological theories on the role of perception and appraisal add to a purely physiological explanation and are closer to showing that the response takes place in a wider context. Yet even though these psychological approaches enhance our understanding of the stress response, they still seriously neglect the broader sociocultural and sociopolitical contexts within which the mechanisms of perception, appraisal, and response occur. At best they show the workings of a personal stress response, but add little to our understanding of how the stress process takes place.

While Selye provided the basis for our understanding of the stress response, his model fails to answer certain other questions. As Pollock observes, Selye's mechanistic approach based on hard biological evidence does not allow for an explanation necessarily based on "the subjective mental processes of perception, evaluation and response" (22, p. 385), and we need to look to other disciplines in order to develop a more complete explanation of the stress phenomenon.

Great difficulties exist in separating physiological from psychological reactions. More recent psychological research has shown the need to account for cognitive and behavioral factors. The work of Monet and Lazarus, for example, shows that appraisal takes place prior to hormonal activity: psychological stimuli precede physiological change that can lead to disease. Psychological research has shown that subtle hormonal responses and subsequent tissue damage can occur through emotional reactions.

We cannot assume that our understanding of the stress concept is complete. According to Mechanic, there are serious difficulties in linking different disciplines in stress research. These difficulties have produced a less than holistic concept of stress. Rather than physiological or psychological responses, the stress process needs to be seen as a psychobiological process, a set of interacting and changing conditions occurring over a period of time.

It is still not known how many hormones act in the stress response and by what combined effect they produce the regulation of physiological processes. However, more recent research has produced sound evidence of limitations to earlier, purely biophysiological concepts of stress, and increasing hormonal research is needed in this area. Some integration of the disciplines (for example, that undertaken by Frankenhaeuser (10)) is essential. Separation of behavioral sciences research from the rest of medical research has meant that much physiological and psychological research have proceeded quite separately.

The stress response must be regarded, at one level, as psychophysiological: a psychological response that has a corresponding physiological response. We cannot carry out adequate model building, however, without understanding the interaction of these processes with broader sociocultural and political processes.

CHAPTER 2

Negative Work Factors and Occupational Stress

Research into occupational stress has developed through biophysiological, psychobiological, psychological, sociopsychological, and sociological approaches. The contextualizing of work factors research into the various roles played by psychological approaches, together with an evaluation of the role played by such research in more meaningful social and structural contexts, is intended to pick up issues of fragmentation in occupational stress research. More recent sociological research has placed much of this watershed of work factors research into a framework in which directions for change in the structuring of work activities and work relations can take place. Occupational stress research is continuing to surge ahead based on the contributions of a number of disciplines; an attempt to place that direction in a suitable perspective may help to identify what is being achieved through new directions. Work relations are seldom included in the language of work factor stress research, yet sociological models offer a way of viewing this research in the context and interplay of control over work relations.

This chapter—a review of the literature on the causes of stress and ill-health— is intended to accomplish a number of objectives. First, it points out the lack of attention to the broader theoretical content of occupational stress: by far the greatest number of studies on the effects of noxious work factors on occupational stress have been conducted by psychologists. Psychological approaches to the study of occupational stress and ill-health tend to be astructural. These approaches are presented and critically evaluated. Second, I will argue that extensive variations occur in conceptual and methodological approaches to the study of occupational stress and ill-health. Despite these differences in approach, occupational conditions have consistently been shown to be predictors of stress and ill-health.

Third, I will review the literature that evaluates negative work experience resulting from lack of control over work processes and lack of opportunity to make decisions at work. This includes the impact of work factors such as lack of control over work and incorporates the effects of both technology and lack of decision-making power. In addition, the effects of a number of other noxious work factors are presented (skill underutilization, role conflict and ambiguity, poor relationships at work, and a negative climate of attitudes from management). There are variations in how lack of control over work has been studied. In fact, as Karasek (65, 66) argues, studies of lack of control represent a broad and separate group of studies from those that show the effects of other noxious work factors. These studies measure an important variable at work that is relevant to understanding the effects of other work stressors.

A consistent theme emerges in this chapter: sociopsychological studies of the effects of work variables almost singularly treat the effects of work factors without considering the broader social and political context within which they arise. I will first review stress studies that adopt psychological or sociopsychological approaches, then discuss those that take labor process approaches. The latter address vital areas ignored by many earlier studies, namely the role of social and structural variables.

A great deal of occupational stress literature has linked stress and ill-health to the experience of negative work factors. The demands of noxious or unpleasant work often can leave an individual's important needs unsatisfied. Coping resources can be important in alleviating the effects of noxious external demands. Mechanic (67) describes coping resources as skills and capacities to deal with external demands: these can enable individuals to exert a level of control over demands. In addition he argues that motivation can be used to meet these demands and that individuals can use psychological defenses to withstand the effects of demands such as negative work pressures. These two components, together with coping resources, are to a great extent the result of social position and hence access to important resources. Whether these are successful or not in alleviating the negative effects of work will influence the person's level of self-esteem and sense of control to effect changes (level of personal mastery). If distress continues and is accompanied by low self-esteem and little sense of personal mastery, ill-health is likely to occur.

PROBLEMS IN METHODS USED TO STUDY
OCCUPATIONAL STRESS

Two elements of the stress process need to be distinguished. One is the stress response, the individual's reaction to negative or unpleasant events; the other is stressors, which are characteristics in the environment that may cause a stress reaction (trouble, bother, or worry). Many studies measure the stress reaction as a response to stressors in the environment.

Spillane (68), a critic of recent approaches to stress research, suggests that many of these studies "consist of uncontrolled observations with poor study design and a lack of critical evaluation in assessing results." One approach to studying stress in organizations is to use subjective measures, asking respondents to indicate their levels of perceived stress through a variety of indicators such as troubles, bothers, and worries resulting from stressful situations. The other approach is to employ objective measures of stress such as adrenaline and cortisol levels, but this also requires participants to give subjective meaning to the objective levels measured.

Frankenhaeuser (10), when speaking of the objective measurement of stress, refers to the psychobiological approach. This approach is based on the concept of arousal. Campbell and Singer (69) discuss some of these approaches. Research has been conducted through urinary analysis of cortisol and catecholamines. When workers are required to achieve high performance levels under difficult conditions, this was found to be done at a cost.

Psychobiological research has certain drawbacks, however. These include the cost to administer such research programs and the fact that subjective measures of associated feelings can give clearer indications of the implications of objective measurements of the stress response. For example, a high adrenaline level in itself may be reported by subjects as a positive feeling (note Frankenhaeuser's work (10)). Research using objective indices, however, has provided valuable data, particularly in relation to the effects of paced technologies, shift work, and overwork.

STRESS AND PSYCHOLOGICAL APPROACHES

By far the greatest number of studies of occupational stress and ill-health have been conducted by psychologists. It is not my purpose here to review all fields of psychological stress research but rather to focus on some limitations adopted, especially by studies that apply psychological reductionism. In fact, there are some great differences among psychological approaches (such as cognitive, individual differences, and personality theories). Some psychological studies suffer the limitations of a lack of a broader social and political perspective and, as such, neglect to place negative work experience within a broader theoretical context.

While cognitive approaches have provided an understanding of a number of interrelationships vis-à-vis stress, there are many opponents of these approaches in psychology. Many writers such as Lazarus (16, 70) and Dohrenwend and Dohrenwend (71) have offered very valuable insights into the process and nature of stress and provide an important starting point in investigating stress. In fact, psychological studies have spanned the range of discipline concerns from biological through to social-psychological issues.

As shown in chapter 1 early approaches to stress research emphasized stress as a purely physiological process. Later developments in stress research led to more recent views of stress as a psychophysiological as well as physiological response. The stress response, however, also presumes an understanding of the emotional and psychological response. This can manifest in physiological reactions. Psychological research has added an extra dimension to the purely physiological approaches, despite its serious limitations in providing a comprehensive framework for understanding stress. Weiss (42), for example, provided one of the few pieces of research that shows the role played by threatening and emotionally difficult situations in producing tissue damage. He showed that the greater the number of unsuccessful coping modes employed by subjects, the greater is the tissue damage. This has important implications for understanding the effects of coping behavior on ill-health. One of the most influential researchers in the modern stress psychological tradition has been Lazarus (70). In some of his early work Lazarus showed the effects of appraisal and coping style on physiological reaction, an approach he has continued to develop (72). He has maintained the theme that emotional response is an important stage in shaping the resulting physiological response in the stress process.

Wassermann (73) is critical of inferences drawn from psychobiological experiments on animals and subsequent inferences about human stress reactions. Numerous studies have been carried out on psychobiological function, emotion, and cognition in the stress response. O'Leary (74) has reviewed research linking emotional response to the immune function. Many other studies have also looked at the effects of these physiological functions.

Reductionist psychological research also has severe limitations. There are many specific problems limiting its explanations of stress. First, sample sizes for research are generally very small. While this is often the result of experimental study designs, it does at times limit the capacity to make inferences about larger populations. In addition, the majority of studies (particularly the more psychiatrically oriented research) deal with problematic populations (groups under threat or some similar pressure) rather than with the behavior and reactions of normal healthy populations. However, the greatest problem is that psychological approaches ignore differences in the social structuring of experience based on class differences. These include differences between the strength and impact of stressors for those in different classes and occupational positions. Reductionist approaches also ignore differences in abilities to deal with stress due to different socialization and different experiences among class groups.

Of the large number of studies of stress in the workplace conducted by industrial psychologists, Nichols (75) argues that they focus on personal characteristics of workers and on environmental factors. Quinlan's (76) review of psychological studies of occupational ill-health and stress shows their lack of concern with broader social and structural explanations of work and ill-health. Quinlan aptly argues that the contribution to occupational ill-health of the social relations of production is not considered within the psychological explanation. There is a denial in industrial psychology of the inherent conflict in the relations of production and the problems it poses. Many of these studies singularly identify symptoms of the conflict inherent between capital and labor in the workplace and treat these as the causes of stress or ill-health. They ignore the existence of power differences in the workplace and different value systems. Psychological studies have largely supported explanations of stress that concentrate on individual characteristics of workers and the role of individual characteristics in the physical work environment.

Quinlan (76) argues that these explanations of stress and occupational ill-health often result in victim-blaming. The most obvious of these studies identify certain personality types as most conducive to stress, and furthermore they ignore the impact of environmental influences. Three outcomes are likely from this kind of research. The first is the recommendation that workers should be selected more carefully for the job. The second is that changes to individuals will bring about a satisfactory match between the person and job requirements. And the third suggests that in order to deal with the cause of stress, at best jobs should be redesigned to suit ergonomic requirements.

Studies of Personality Types

Numerous studies have been conducted by psychologists on the role of personality dispositions in the experience of stress. Their principal research interest has focused on two types of personality behavior patterns, Type A and Type B. These studies have identified Type A as hard driving and competitive with a strong achievement orientation. According to Sutherland and Cooper, Type A personalities also have "a strong sense of time urgency, hurried and explosive speech patterns, quick motor movements [and are] aggressive, hostile, restless and impatient" (77, p. 74). Type B personalities are reported to be more relaxed and to not display Type A characteristics. Dunbar (78) was one of the first psychologists to describe these personality type characteristics.

A major study of the impact of personality type was the Western Collaborative Group Study conducted over eight and one-half years and reported by Rosenman and associates (79). It showed that Type A personality employees were twice as likely as Type B to develop coronary heart disease (CHD). The Framington Heart Study (80) also supported this association: however, neither study, according to Sutherland and Cooper (77), found a personality–CHD risk association for blue-collar workers. A review of the literature on personality type and stress by Matthews and Glass (81) concluded that when confronted with uncontrollable events, Type A personalities will attempt to gain control by using more effort than Type B personalities. Ham (82), however, in a recent study of women and stress found no significant relationship between personality type and stress, and concluded that Type A personality is as much a coping strategy as a fixed personality characteristic.

Type A personalities have been found most likely to experience stress and ill-health. Caplan and coworkers report "significant occupational differences in the Type A coronary-prone personality" (83, p. 202), with administrative professors, administrators, physicians, and blue-collar tool and die makers displaying the highest instances of this personality type. However, their study concluded that "there was no clear evidence that personality variables directly affect psychological, physiological, and behavioural strains."

Lamude and Scudder (84) found that while Type A personality managers did not tend to use flexible decision-making, avoidance conflict was not related to Type A. Denolette and de Potter (85) reported that certain coping styles, stress, and Type A were related to other factors also. Greenglass and Burke (86) found that work stress was moderated by coping for Type A personalities only, while Byrne and Reinhart (87) found job satisfaction to mediate the effects of Type A.

Jamal (88) found Type A to be an important moderator of stress-outcome relationships. Davilla and coworkers (89) reported that Type A personalities experience specific stress symptoms more often than Type B. Weiss (90) in an East German study reported level of social status, cholesterol, and high blood pressure among other factors correlated with Type A. Type A effects were also found by Nahavandi and associates (91) and Barnes (92).

While a number of studies show Type A stress associations, evidence for the effects of personality type on stress is inconclusive. There is some evidence that Type A personality is related to CHD. For example, as early as 1948 Dunbar (93) cited evidence, anecdotal or otherwise, to suggest a relationship between personality type and CHD. However, Deary and coworkers (94) found other psychological factors had stronger effects on hypertension and CHD than did Type A. Also, certain components of Type A responses such as anxiety (e.g., 95–98) and depression (e.g., 95, 99, 100) have been found to be related to stress outcomes. Caplan and associates (83), Karasek (65, 66), and others utilize these outcome measures in studies of stress at work. It is likely that, following Ham (82), Type A identifies a coping mode for dealing with work stress.

A problem is that the relationship between variables in stress is more complex than the action of an individual personality variable. It may well be an important interactive variable, but just how it fits in a stressor, stress response, and coping style model is not well understood: it needs to be studied in interaction with a number of other variables in the overall stress process. A problem with Type A analysis is that conservative interpretations of the implications can lead to recommending change in individuals rather than looking at social, structural, and other variables. It may lead to changes that favor managerial rather than employee needs and interests.

RESEARCH INTO NEGATIVE WORK FACTORS

The study of occupational stress prior to the 1970s tended to be based on an analysis of interpersonal relations and psychological factors. While these factors still predominate, more recent research has looked at the effects of noxious work demands. A considerable amount of research has been conducted on stressors in the workplace; while much of this research has been conducted by psychologists, it does identify the effects of alienating working conditions on stress. More importantly, it helps establish a link between the presence of alienating working conditions, stress, and ill-health.

More recent research has identified the role of perceived control in affecting the impact of negative work factors on stress: it may also produce stressfulness at work. Karasek (65, 66), Israel and coworkers (101), Andries and coworkers (102), Hall (103), and a number of labor process writers (e.g., 4, 5) are among those who identify the importance of a lack of control at work.

Research into occupational stress has covered a wide range of areas, from the effects of technology, particularly mass production and machine-paced work and the effects of noxious and dangerous environments, to the impact of work factors such as poor relationships, role conflict, and ambiguity. One common conclusion from these approaches is that detrimental work environments and conditions can have sociopsychological consequences and possibly ill-health consequences for workers at all levels, especially at lower levels.

Levi (17) argues that four broad job factors are critical to an analysis of occupational stress: qualitative overload, quantitative overload, lack of control, and lack of social support. Cooper and Marshall (104) extend the list somewhat. They add factors intrinsic to the job, such as physical conditions including danger, organizational role encompassing ambiguity and conflict, and career development. In addition, relationships at work and organizational structure and climate are important causes of stress at work. Caplan and associates (83) in their study of health and job demands also use job complexity and the underutilization of skills and activities. Physical working conditions often are not included in a discussion of occupational stress. Levi (17) argues that a trend in research has been a distinction between those studies that look at psychosocial factors as the basis of stress and those that look at physical work conditions.

With the exception of Levi's work, few studies establish a broader context for the presence of noxious working conditions. As such they explain little about the characteristics likely to lead to the experience of stressors. These studies contribute more to an understanding of the causes of stress than purely physiological and psychophysiological studies. The sociopsychological studies of occupational stress identify characteristics in the environment likely to lead to stress; psychophysiological studies described characteristics of the individual stress response. Despite some important drawbacks, these sociopsychological studies come closer to providing a fuller explanation of the process of stress. They do not, however, provide an adequate account of the broader context within which occupational stress takes place. They also fail to account for differing social, cultural, economic, and political demands and resources across different groups in the community. As such, they tend to be astructural.

While studies of the effects of lack of control such as those by Karasek (65, 66) address its important mediating effects, they tend to ignore questions of who exerts control, how it is established and enforced, and the system of work relations that it helps to maintain. Studies in the tradition of the labor process in the field of sociology take up these issues.

The following review covers a range of factors present at work that have been treated in previous research as the basis of stress investigations. Some factors have not been included, either because at this stage insufficient research has been conducted or because they do not provide a basis for comparative analysis across occupational groups in the study. Factors included are: control over work, lack of skill discretion (including underutilization and routine work), and role ambiguity and conflict. In addition, the effects of poor work relationships, negative attitudes toward workers by management, and work overload are reviewed.

Control Over Work

A fairly substantial number of writers have argued that lack of control at work is a major source of stress and consequent ill-health. Karasek (66) describes two traditions in stress research: the first looks at decision latitude and the second at the consequences of other work stressors. He argues that the first type of study deals with how the work itself is carried out: the measure most often used is a perceived lack of control to influence the job. The second tradition deals with other work factors, aspects of the structure of work organization, and the conditions under which employees work.

A Perceived Lack of Control. Many studies have been conducted on the effects of a perceived lack of control on stress and ill-health. The variety of measures of lack of control is only surpassed by the range of work, stress, and health measures to which they are related. Three principal types of studies evaluate the effects of lack of control: those that look at the interaction between control and work demands (e.g., 65, 66, 105–107), those that look at the effects of machine pacing (e.g., 108–110), and those that consider a lack of control over work process (e.g., 101, 111, 112).

In 1981 Karasek (66) conducted a major study of job discretion in relation to work demands. He and others have employed lack of control to exert influence as part of a wider notion of a lack of discretion at work. In a study of both Swedish and U.S. national populations he found that ability to exert control was coupled with the utilization of skills and job variety as measures of influence and discretion at work. His model measures decision latitude against job demands. Karasek found that a low level of decision latitude compared with large job demands increases job strain and leads to poor mental health. Any jobs with insufficient decision latitude have the poorest mental health outcomes regardless of the level of demand. Karasek found that jobs lower down in the hierarchy are characterized by least decision latitude, including decision authority, and little opportunity to exercise control, use skills, and perform varied tasks.

Karasek argues that job decision latitude "measures the working individual's control over the use of skills on the job and his authority to make decisions about the organisation of his work. . . . Another aspect of decision latitude is the worker's ability to exert control over uncertainties at work" (66, pp. 79–80). He found a high correlation between job discretion and autonomy, and thus included them in a single measurement scale. In his study he argues that increased decision latitude is preferable, rather than reduced job demand, in reducing mental strain. Conceptualization of measures in the study allowed for a multidimensional analysis of task-related utilization and opportunity for control. According to Karasek (66), by increasing decision latitude, both strain associated with workload and the problem of passivity at work would be alleviated: this would allow for increased actions and increased skill utilization on the job.

Warr (113) tested relationships between decision latitude and job demands, while two recent studies of the work demands–job decision latitude model by Perrewe and Ganster (105) and Landsbergis (106) support Karasek's conclusions. Several somatic outcomes were found to result from insufficient scope to deal with the demands at work. Braun and Hollander (114) found depression an important outcome for women. Occupations with high decision latitude were shown to have relatively low cardiovascular risk (115).

The second type of study on the effects of lack of control on stress considers the effects of machine packaging. Frankenhaeuser carried out a large number of studies using indices such as adrenaline, noradrenaline, and cortisol levels. Frankenhaeuser and Johansson (108) found that work conditions that stimulate an ability to exert influence and control will reduce the neuroendocrine stress response and increase job satisfaction and well-being. Broadbent and Gath (110), Frankenhaeuser and Gardell (109), Smith and associates (116, 117), Wilkes and associates (118), and others have shown harmful effects of machine pacing. Hurrell and Colligan (119) warn, however, about the need to account for individual and situational determinants of the deleterious effects of machine pacing. Bulat (120) has suggested ways of improving paced work, while Franks and Sury (121), Manenica (122), Cox (123), and Johansson (124) have also shown that increased stress occurs with paced work. Hokanson and colleagues (125), Johansson (126), Kornhauser (127), Caplan and colleagues (128), and Savery (129) have also reported the adverse psychological and physiological effects of machine-paced work.

The third type of study shows lack of control over work as an important factor in considering the effects of other sources of strain. Karmaus (111) assigned a key role to decision latitude in the interaction of factors that contribute to stress and ill-health. Other studies (e.g., 101) found that satisfaction with influence had significant effects on stress and job strain; Nelkin (112) also found perception of control to be important. Caplan and colleagues (83) reported the amount of decision influence to be lowest in lower occupational groups.

Aronsson (130) suggests a number of levels of control, many of which have been incorporated in the previous analysis, but with qualitative differences. First is the type of control in which workers can affect an outcome. Control can also be conceptualized as predictability, as participation, and as control that can affect change over the governing rules (a distinction from control within situations). Finally, he adds that a distinction should be drawn between collective and individual control.

In an evaluation of the relationship between production control and chronic stress, Houben (131) has identified five types of production control: direct, division of work and resources, socialization, sanctioning, and power development. He argues that this typology is a way of identifying several sources of control and their effects on chronic stress.

Despite some wide variations in types of measures used, few of these studies have located lack of control at work in a broader context of social and political processes. The majority of studies ignore issues of differences in the opportunity to exercise control among different class groups.

Relationship to the means of production is an important concept in the determination of social class (e.g., 132, 133). Certain studies have examined the relationship between social class and control: these studies tend to operationalize control as an important variable. Kohn and coworkers (133) examined the relationship between social class and the experience of distress in three different countries and found that while factory and nonproduction workers in Poland reported less distress than higher classes, in the United States and Japan the pattern was reversed. Other studies such as that by Otto (134) examined a number of selected occupational groups in Australia and found that differences in stress demonstrated class differences. Other studies of stress differences among occupational groups include that of Anderson and coworkers (135). Relatively few analyses of class differences in stress have been carried out. Variations in the effects of social class on stress were found in a U.S. sample by Barney and associates (136). Schwalbe and Staples (7) also refer to class in their analysis of occupational differences in stress. Unskilled blue-collar groups reported the greatest stress.

Decision-Making Power. Levi (17) argues that as factories grow, so does the distance of workers from both management and consumers. In addition he argues (17, p. 18):

> mass production normally involves not just a pronounced fragmentation of the work process but also a decrease in worker control of this, partly because work organisation, work content and pace are determined by the machine system, partly as a result of the detailed pre-planning that is necessary in such systems. All this usually results in monotony, isolation, lack of freedom and time pressure, with possible long term effects on health and well-being.

According to Israel and coworkers there are "two main categories into which the concept of control has been operationalised" (101, p. 165). One is employees' control over the job and the other is the "degree to which employees are able to 'participate in decision making'."

Studies by Frankenhaeuser and colleagues (10, 137) showed negative effects of a lack of control, while French and Caplan (138) present a model of participation and occupational experience. Gardell (139) and French and Caplan (140) report high satisfaction and esteem and lower job threat from greater participation. In Australia as in a number of other countries, the union movement was slow to respond to government initiatives on worker participation; its opposition to team working models was also quite strong. These initiatives have taken some time to gain acceptance because of perceived demarcation problems and pay threats as well as employment insecurities. Participation initiatives have not readily achieved their goals.

Emery and Thorsrud (141) evaluated autonomy in relation to productivity. Gardell (139) and Jackson (142) found participation to affect perceived influence. Cooper and Marshall argue that "research seems to indicate that greater participation leads to lower staff turnover, higher productivity, high performance improvements . . . and that when participation is absent, lower job satisfaction and higher levels of physical and mental health risks result" (104, p. 22). Kommer (143) found that increased autonomy can alleviate stress. Evidence demonstrates a clear link between participation and certain health indices, particularly levels of mental health and strain.

Other Negative Work Factors

Karasek (65, 66) found that when there was little skill discretion, measured as a lack of opportunity to use skills and little variety on the job among other measures, the effects of work demands on mental ill-health were greatest. Little skill discretion combines with a number of factors such as discretion and task variety to cover latitude of influence. While O'Brien and coworkers (144) found little relationship between underutilization and somatic ill-health, Spillane (68, p. 594), Van der Auwera (145), and Lowe and Northcott (146) found negative attitudes, distress, and ill-health resulting from underutilization. Frankenhaeuser and Gardell (109) and Caplan and associates (83) conducted research into the nature of underload, particularly in relation to overload.

A number of writers have demonstrated difficulties associated with role-related problems. Kahn and associates (147) conducted the first study into role ambiguity. French and Caplan (140) and Martin (148) reported on CHD and related ill-health outcomes, while Travers and Cooper (149) found resulting mental ill-health. Hammer and Tosi (150) and Savery (129) studied resulting job dissatisfaction. Kahn (151), French and Caplan (138), and Sutherland and colleagues (152, 153) reported on ambiguity, tension, conflict, and role stress.

Few studies have examined the effects of the perception of control by management on the organization of work. Leigh and coworkers (154), however, studied perceived control by management and role conflict. Much research into role-related problems is fairly inconsequential in that reported relationships between problems and health measures are not strong. However, enough significant results have been produced to show that job dissatisfaction and tension and ill-health result from role conflict.

While poor relationships at work have proved to be a cause of stress, few studies have treated poor working relationships as an outcome of employees' lack of control over work processes and work environment. French and Caplan (138), Kahn and associates (147), Guppy and Gutteridge (155), and Cooper and Marshall (156) found that inadequate communication, poor interpersonal relations, and less than effective social support led to stress. Caplan and associates (128) and van Dijkhuizen (157) found that peer support reduced stress. Van Dijkhuizen (157) and Buck (158) recognize that poor relationships, especially with management, can lead to stress. However, much work on stress and the quality of relationships at work relies on attitudinal measures and little has been related to indices of health. Work relationships interact with many other variables at work, including the technological requirements and demands of the job, the degree of support good relationships can provide, and occupational position. More systematic research is needed in this area.

Certain studies have demonstrated the effects of management's organization of work on stress. Again, however, few of these studies conceptualize this measure as part of a broader problem of lack of control by employees. Much research in relation to stress outcomes has focused on the effects of job factors (such as work constraints, control, supervision) and their effects on other psychological and personality dimensions. Some Scandinavian research (e.g., 159) has focused on the effects of broader organizational values such as democratic versus autocratic. Cross-cultural differences in the effects of organizational structure and climate were found by Kircaldy and Cooper (160). The majority of research has operationalized control at the task or organizational structure and procedural level: debates on factors such as the effects of machine pacing, nature of supervision, and effects of workload have had a predominant effect. More recent writings by Child (161), Littler and Salaman (162), and others on the nature and extent of control beyond the process of production have largely been neglected. Much of the debate on control has centered around established territories. Nelkin (112) and Masreliez (163) describe the effects of negativity or no feedback from management, while Bolinder (159) records negative effects of distinctions made by management between blue- and white-collar workers. An autocratic style of management control and managerial dominance are likely to show in negative attitudes, especially toward lower level employees: stress is likely to occur through constant negative opinions.

McLean argues that "the pressure of having too much to do or at least feeling that one has too much to do would seem to be a fairly obvious stressor" (164, p. 81). French and Caplan (140), Kornhauser (127), Frankenhaeuser (10), Sutherland and Davidson (152), and Siegrist and Klein (165) report negative effects on health from qualitative overload. Studies by Endresen and associates (166) and by Richardsen and Burke (167) on medical doctors found workload to have the highest effect of a number of stressors. The effects on alienation have also been reported (168). For Petitone (169) the perception of overload was important. Broadbent and Gath (110) argue that underload is becoming a source of stress as technologies become more advanced. A number of researchers have found confounding effects of other variables for overload. Breslow and Buell (170), Kirmeyer and Dougherty (171), Rissler (172), Karasek (65, 66), and Bruner and Cooper (173) have variously found that age, sex, occupational level, decision latitude, and corporate performance have mediating effects on overload.

Garfield (5) is one of the few researchers to argue that overwork and underutilization can also be understood as an outcome of hierarchical structures and management motivational techniques. Increased organizational responsibility for administrative and supervisory employees in a Canadian school district led to decreases in stress, even though the intensity of other stressors increased. While writers such as Ratsoy and coworkers (174) found work load a major cause of stress, Del'Erba and coworkers (175) did not confirm this association.

Research into each of the negative work factors included here shows them to have some important effects on stress. These studies, however, have largely ignored a sociopolitical context for the presence of these stressors and for reasons why certain groups of workers would experience them as stressful. A limitation of many of these studies is that they only describe the presence of a stressor, rather than providing a context within which the negative work factor's presence can be explained as part of a wider social structural context.

Other Sociopsychological Factors

A number of studies have examined the effects of work factors on self-esteem and their consequent effects on occupational ill-health. Again, many of these studies do not treat low self-esteem as an outcome of the structure of work relations in which employees experience little control. The work of Kohn and associates (133) points out the effects of class membership in the United States, Japan, and Poland on degree of self-direction. Those more advantaged were found to both value and experience self-direction. Staples and associates argue that in a Marxist sense class position "either constrains or enables the experience of self-efficiency by structuring one's activity in the work place. . . . A considerable amount of previous research has attempted to establish a relationship between position within the stratification system and self evaluation" (176, p. 89). They conclude that social stratification affects the way in which self-esteem is formed. While Schwalbe (177) examined the sources of self-esteem at work, Pearlin and coworkers (178), Israel and coworkers (101), and Margolis and Kroes (179) all reported negative effects of low self-esteem.

Pearl and colleagues (178) and Israel and colleagues (101) reported on the sociopsychological resource of mastery, another work outcome, in order to explain poor health. Locus of control has been associated with physical and mental health by Kircaldy and Cooper (160), Rees and Cooper (180), and Daniels and Guppy (181). Relatively few occupational studies of stress and ill-health, however, have included mastery as an independent or mediating variable in the stress response. Nonetheless, low self-esteem and low personal mastery have been found to be influenced by negative work factors. With the exception of Schwalbe (167) and Staples and coworkers (176), authors have treated them more as individual responses outside a broader social context.

The Effects of Ascribed, Achieved Status and External Sources of Stress

A large number of factors contribute to stress. Many of these are related to variables that exist external to and before exposure to noxious work factors. The effects of age have been referred to in a number of studies; Osipow and associates (182), for example, showed its effects on stress, while stress has been associated with "feeling" older than one's chronological age (183). Hinkle and associates (184) and Wan (185) found a relationship between level of education, occupational position, and CHD morbidity. Many other studies have reported these variables.

Some studies have included extra-organizational factors in their evaluations of work stress. Home life, for example, can have some positive benefits on stress experienced at work. Studies of work and nonwork stress, including home responsibilities, have been reported by Kleitzman and coworkers (186), Bolger and coworkers (187), Hall (188), and Burke (189). Spillover effects have been found from both home stress on work and work stress on the home (190, 191). German managers found the home/work interface a greater source of stress than did British managers (160). A number of other factors can come into effect, such as whether one is in a domestic partnership and, if so, whether the partner is employed, and the number and ages of children. Avison (192) found that the relationship between family structure and employment had an effect on women's stress. Gill and Hibbins (193) report that family role contributes to the stress of wives. The friendships made at work and the opportunity for independence can be flow-on benefits to the home. While dual-role stress has been reported by working mothers, some important benefits were gained from working (194).

However, very few studies have investigated the interactive effects of work and home life.

BLUE-COLLAR, WHITE-COLLAR, AND MANAGERIAL STRESS

McLean argues that "research into stressors at work bears out the notion that literally anything can be termed stressful if the individual's vulnerability is extraordinarily high and if a supportive environment is unavailable" (164, p. 73). Blue-collar workers have least control, least opportunity for influence, and least support from management and structures. McLean suggests that certain "occupations have been associated with greater risk of ulcers, coronary heart disease, and purely psychiatric disability," and argues further that similar associations "can be made between illnesses and social class and specific ages" (164, p. 73).

Shift work, changes, monotony, and conflicting work requirements or ambiguous work contribute to stressful reactions. McLean (164) contends that job loss or the threat of losing one's job are more potent stressors: that the growing threat of unemployment is a detrimental stressor that has negative effects on health. Shostack (195) argues that it is a more constant threat with greatest stress and ill-health implications for blue-collar workers. He further suggests that blue-collar workers have several characteristics that shield them from the harsh realities of low paying jobs with little satisfaction: they are able, through "self-esteem protection [to] . . . maintain a positive notion" of providing adequately for a family (195, p. 12). He also argues that the inflationary rate adds to "prop up the illusion of economic well-being" despite comparatively low incomes, and suggests that (195, p. 12):

> male blue-collarities are frequently shielded from the full story of their own families economic stringency by wives who want to spare them "additional worries.". . . Male blue-collarities learn from decades of experience to mute and even deny dissatisfaction with stressors that seem out of their hands that they have little clear hope of reforming.

These, in Shostack's view, help explain the regular underreporting of the effects of negative work experience and, as he points out, several national U.S. studies have shown that blue-collar workers' health levels are not only poor but are most often caused by work-related problems. The physical working conditions of dirt, dust, noise, and dangerous chemicals and solvents can cause a large number of acute and chronic symptoms. As Shostack claims (195, p. 27):

> substantial disparities between [blue-collar workers] neglected work station and the pampered front office . . . in the noise quality levels, standards of cleanliness, general disorder, and even aesthetics (as well as temperature

variations, inadequate ventilation and air conditioning) . . . are taken by many as a stressful measure of disdain.

Several different types of negative work outcomes are characteristic of blue-collar work. Shostack argues that a low status derived from this type of work can lead to self-denigration. In addition, work rules and production pressure can severely diminish freedom on the job and can lead to harassment. This in turn can lead to feelings of not being appreciated and of not having other important needs met. Finally, Shostack argues that the quality of work relationships is likely to be poor: "blue collar camaraderie . . . appears to be a source of strain" (195, p. 56).

Shirom and Kirmeyer (196) treat coping as a mediating factor in the relationship between stress and somatic complaint. Coping involves activities that reduce the harmful effects of stress. A survey of 251 employees compared the effects of union membership and nonmembership on outcomes of somatic complaint. Union members reported higher levels of role and interrole conflict and ambiguity than nonmembers but similar levels of somatic complaint, demonstrating how union membership as a coping strategy modifies the outcomes from work stress. Stress levels were also affected. The more effective the perceived performance of the union, the lower was the level of perceived stress.

The incidence of somatic complaint has been related to class membership in a number of studies, such as those of Garfield (5) and Coburn (4). Lower social classes have limited resources to deal with the effects of sociopsychological work characteristics. Medical risk factors are distributed more frequently toward the lower end of the social class continuum and can be related to psychosocial work factors. Marmot and Theorell (197) point out that cardiovascular illness and its differential occurrence may be due to psychosocial work differences. They found that these differences are likely to have neuroendocrine consequences and effects on lifestyle.

A great deal of the research on the effects of negative work factors reported here has shown that blue-collar workers, those groups with the least control and the most negative work experience, have the greatest stress and lowest levels of health. Kornhauser (127) found that noxious working conditions and the need to work hard contributed to poor mental health: this was found to be poorest in the low-status jobs. Schabroeck and coworkers (198) found that participation and coworker support had important effects on role conflict, ambiguity, and satisfaction. However, programs have been tested for abolishing strain among blue-collar workers (199). McLean argues that "previous research [concludes] that the lower ends of the socio-economic continuum are more prone to stress reactions than those in the middle or upper classes" (164, p. 80).

White-collar office (clerical and professional) work has suffered some of the effects of deskilling and job fragmentation due to increased management control and increased mechanization. However, white-collar workers tend to be middle class and aspire to and generally support the values and ethos of management.

Except for lower level, clerical and support (secretarial) staff, the expected career path for many white-collar workers could be into management. Their physical working conditions have more in common with management than with blue-collar work. In fact white-collar workers tend to represent far more the interests of management than the interests of working-class occupational groups. There are some exceptions, but class similarity to management means that they share many similar stressors.

Lower level office (secretarial) work is often affected by the demands of technology, as in the case of work with visual display units (VDUs) and other computer operating jobs (200). Arora (201) has studied the effects of supervision and related factors on the alienation and physical health of VDU and non-VDU users. Work can also be fragmented, offering limitations in skill utilization. Most white-collar work enjoys higher status and opportunities than blue-collar work, and fewer stressors. Many white-collar jobs however are classified as "women's work." Management control affects the work of different occupational groups. According to Balshem (202), 73 percent of more than 400 female clerical workers in a U.S. study found stress a characteristic of their job. The workers who reported their bosses to be inconsistent and emotionally harassing found this a major cause of job stress.

A number of factors have accounted for stress among white-collar workers. Turnage and Speilberger (203) reported that lack of opportunity for advancement and type of supervision accounted for higher levels of stress. Clinical staff were found to suffer the effects of a large range of stressors, more than managerial and professional staff (204). Decision latitude was found to have a moderating effect on the levels of stress for clerical staff, rather than for managers (205). In addition, greater stress has been associated with lower satisfaction for white-collar employees (206).

Research on managerial stress has largely looked at the acceptance of responsibility and qualitative and quantitative work load as the primary causes of stress. Marshall and Cooper (207) found that work overload was a primary cause of stress at work. Many other causes are related to an interaction between work life, home, and leisure.

Several sources of stress for management are suggested by Moss (208). These include pressures from the process of management, and the increased rate of change in business. In addition, Moss argues, there are new "challenging established organisation practices, . . . pressures from the manager's personal life such as family and social goals and responsibilities, . . . [and] the changes in self-image, perspective, and capacities" that go with changes in adult life (208, p. 6). I maintain further that with the growth of large and multinational corporate life and highly competitive business environments, business life can develop a sense of powerlessness that may also affect some management roles.

Managers, whether senior managers, middle management, or supervisors, have particular expectations about role performance. Often these are self-imposed as they can affect promotion prospects and career development. Moss (208) argues that personal problems can often be seen as reflecting low competence. Managers may not wish to recognize problems and difficulties. At worst they are likely to keep problems to themselves. In addition, Moss argues, they may suppress their anger at being treated poorly by their supervisors due to "fear of retaliation," and "feelings of apprehension and a desire to call for help in response to unexpected contingencies or emergencies may be considered by managers to be signs of weakness or dependence " (208, p. 8).

Apart from problems due to home/work interactions and pressure, Moss suggests several other sources of stress for managers. Relationships with superiors and subordinates, the impact of work pressure transmitted downward from the top, problems with too much or too little responsibility, and worries about future career prospects may all be sources of stress. Moss argues, following Minzberg (209), that managers work unrelentingly at great quantities of work. The manager's sense of competence and self-worth is dependent on being needed and having a valued role to play. Central to this is the emphasis placed on career development and a lack of job security, fear of obsolescence or redundancy, and underpromotion. Part of the ideology of managerial work, however, is that pressure is not necessarily bad, although pressure experienced by managers who have substantial control will have different consequences from those for managers with little control.

The role of lower and middle management has several sources of pressure not associated with senior positions. Moss (208) argues that middle managers receive little consideration or support from the managers above them and the workers below them. Managers are not exempt from the deleterious effects of powerlessness and a lack of control, even though they usually have greater opportunities than other employees to exert influence in the organization. Rather than being in substantial control, too many managers spend their time reacting to the demands of the chief executive or board of directors. Das (210) found that powerlessness induced by organizational arrangements and too much risk-taking were associated with high stress. Power, support, and the nature of communication networks were found to contribute to high stress. A study by Marshall and Cooper (207) of 200 British managers examined the physical and psychological outcomes that could be related to a number of conditions including work. They found that even for senior managers, a lack of autonomy played an important role particularly in the incidence of anxiety. For managers in production settings, physical health was predicted mostly by a lack of achievement while psychological health was affected by a lack of job security, work overload in a challenging job, and a lack of autonomy (experienced only in large companies).

Lack of organizational support was found to be related to higher stress for managers as well as white-collar workers (203). Westman (205) argues that managers have more coping resources related to their position of power to mediate the effects of stress than do white-collar employees.

Associations between high stress levels and somatic complaints, especially cardiovascular disease, have been found for managers. Associations between health outcomes and high stress are consistently presented in research. Moss cites a causal relationship between CHD and managerial status: that is, where job requirements are ambiguous, there is a great responsibility for people, work demands are conflicting, and the manager feels compelled to act against values or beliefs. He argues further that the incidence of CHD can be related to the need to strive for personal and organization goals: "work pressures and striving are perceived as constructive forces" (208, p. 71).

Research has established that the experience of stress and ill-health differs among occupational groups. I have shown that a lack of control over the labor process, and therefore a lack of control over the conditions under which work is carried out, is an important factor in determining stress and ill-health. No occupational group is exempt from experiencing stress and its consequences; however, groups that experience least control over work are likely to have the greatest stress consequences and the poorest health levels.

CONCLUSION

Of the studies reviewed, some contribute to an individualistic notion of the causes of stress and focus on characteristics of workers and their propensity toward stress. The broader stream of studies focus on the nature of work and working conditions. Yet these studies are limited in explaining the nature of the causes of stress at work and lead only to piecemeal solutions. There are few studies that relate the nature of work, working conditions, and management policies and practices to the labor process. These few studies view stress as emanating from working conditions that result from competition for control over the labor process.

Social class has been shown in a number of studies to have important effects on stress (e.g., 133). Relationships to the means of production are important in determining relationships to control. The production of surplus value by those with little control and the benefits gained by those with large spheres of control (owners and managers) from the organization of production relations are precursors for the experience of stress. A number of studies (e.g., 83, 133, 134) have shown control over production relations to be less conducive to stress than little control (or autonomy). Issues of control and associated greater stress for those being controlled are important in an analysis of the effects of class differentials.

CHAPTER 3

Sociological Approaches to Stress and Labor Process Theory

Sociological approaches provide a most fruitful way of understanding the process of stress and other sociopsychological states. Earlier physiological, psychophysiological, and later psychological approaches, reviewed in the previous chapters, have identified aspects of stress; however, they neglect the impact of important contextual factors. Treating stress as an outcome of broader social, cultural, and political forces is the most fruitful way of understanding its causes and of understanding stress as a process. Stress can also be understood as a process rooted in structural causes and resulting in identifiable patterns of individual experience.

Many different sociological perspectives are used in industrial and medical sociology and in the sociology of occupational health and safety. Not all sociological theory relating to health and illness at work has been critical, however. For example, not until relatively recently did industrial sociology adopt a critical perspective: earlier it had been dominated by a preoccupation with workgroup design and group process, issues that interested psychologists.

A broader sociological framework based on a labor process approach is introduced here to explain occupational stress and ill-health: that is, stress and ill-health result from broader social processes. This means that stress will be viewed as an outcome of management's control of the labor process. In Chapter 4 I will examine more recent developments in the labor process approach and its implications for occupational stress and ill-health and will address the importance of alienation as a concept that helps to explain occupational stress. Chapter 5 then covers a number of studies that treat stress in its broader socioeconomic and political context. The role of government policy in helping to shape the labor process and in intervening in stress and ill-health outcomes of work will also be presented.

Very few writers discuss stress in a wider economic framework or in the context of production relations in society. In addition, management's control of the labor process in a capitalist society can be seen as fundamental to explaining the causes of occupational stress and ill-health for workers. Sociologists arguing from a Marxist position have shown that class relations, and the domination by those who own and manage of those with only labor power to sell, form the backdrop for understanding labor management relations. This creates large divisions between management and workers (see Blauner (211)). Experiences of different occupational groups are deeply rooted in the relationship of ownership and control and the social relationships that grow out of this.

DIFFERENCES IN WORK EXPERIENCE

There are many explanations for differences in work experience and outcome between managers, other white-collar occupations, and blue-collar groups. Management literature has strongly emphasized that work organization results from the need to achieve productivity and increased growth. While the Gilbeths (cited in 211) and other early writers in management constructed classical principles of the "best" way of managing organizations, it was Frederick Taylor (3) who was one of the first to prescribe severe organizational strategies for workers to ensure that management could fulfill production goals. The specific effects of Taylorism on work organization have been described by many writers (212–214).

Sociologists, however, have identified these explanations of work organization as management based and as a justification of management control rather than a recipe for industrial success (4, 215). In modern western capitalist societies, a rhetoric has developed among industry and business leaders as a way of explaining management's imperative to control: the rhetoric states that the need for productivity, growth, and survival occurs in an environment of increasing international competitiveness. The cost to workers, however, is great. In many countries the rates of injury and illness are high (216), and in smaller countries such as Australia there has been a widespread increase in the extent and cost of occupational health and safety incidents (217).

With increased pressure to compete and focus on productivity increases, the opportunity for workers to meet important needs at work diminishes. And this is of secondary importance to most employers, management, and government, with high levels of unemployment forcing employees to retain their jobs even while conditions of work deteriorate. The chief concern of management and governments is the fulfillment of the capitalist objectives of profitability and productivity.

The overview of the labor process in this chapter will identify two major impacts of the capitalist labor process: first, the increasing levels of management control over that process, and second, the reduction of skill levels in many traditional occupational groups. In Chapter 4 I will look at the role of technological development as a means of management gaining control over the labor process and further alienating workers.

SOCIOLOGICAL RESEARCH:
ITS CONTRIBUTIONS TO THE FIELD OF STRESS

The sociological approach offers a broader explanation of the process of stress than do physiological and psychological approaches. This is not to deny the very important contributions that the latter disciplines have made to understanding the biophysiological mechanisms involved in the process of stress, and the individual perceptual and cognitive elements of the process. Sociological approaches have addressed environmental, cultural, and social contexts and thereby offer a fuller evaluation of why stress takes place, beyond biophysiological response and individual perception. Apart from identifying characteristics of the environment within which stress occurs, these approaches provide a framework for interpreting population patterns and trends in stress. Based on the work of Marx (218), Weber (219), and Wright (220), for example, a basis of social class differences can be argued.

Sociological approaches account for sociopsychological experience in relation to stress, emphasizing socialization, perception, and interpretations of situations. Differences between individuals have been dealt with in relation to differences in social class. However, very few studies have actually addressed sociopsychological effects of stress in different occupational groups at work (see Otto (134), Kohn and colleagues (133), Schwalbe and Staples (7)). The effects of social structural factors in the occurrence of stress, however, are an important part of the sociological domain.

A number of fields of sociological research have addressed stress at work. Two are of primary concern here: interactionist sociologists address individual perceptual processes in the stress response that result from social and cultural experience; sociostructural sociologists focus on the impact of social, political, and cultural structures on stress. There are other approaches to stress in sociology, but these two are the major domains of interest.

Interactionist Approaches: Definition of the Situation

Interactionist sociologists focus on individual sociopsychological responses to social phenomena and are interested in the way that people construct world views based on their social and psychological experience. Thomas (221) and Cooley (222) have developed arguments to show why people are essentially cognitive in their actions. This is important because it gives meaning to experience and enables people to define situations and to act on the interpretations and meanings they give. Thomas (in Volkart (223)) and Cooley (222) focus on how situations are perceived and demonstrate that they are determined by social experience. In interactionist sociology, it is characteristic to focus on defining the situation and the action that takes place as a result of it (224).

This distinction is useful for explaining stress as it highlights that action based on stress is tied to perception. This is not the only interpretation of behavior as a response to stress. If an event is defined as negative, the response will draw on actions and behavior that focus on negative attributes in the situation. This approach does not attempt to deny that some external stressors will affect action, irrespective of cognition of stress (see Chapter 1), but it does highlight the important role of perception and definition in the type of response to stress.

Thomas argues that the meanings of objects are important "because it is these meanings which determine the individual's behaviour" (in 223, p. 156). Mechanic argues that Thomas's concept of crisis contributes in two main ways to understanding the stress response. According to Mechanic, Thomas shows that adjustment and control depend on the "preparation of the individual to meet events" and that crisis does not lie in a situation but in the "person's capacity to meet it" (63, p. 293).

This approach was recognized quite early in stress research, with Wolff (225) arguing that stress was due to the way in which individuals perceive a situation. He further argued that genetics, individual needs, conditioning, and cultural and general life pressures help to shape the way in which a situation is perceived. The past of the individual is an important factor in disease etiology. Two people may be affected quite differently by the same stimulus. This identifies differences in the physiological and psychological responses resulting from perceiving a situation as threatening or noxious. Wolff concludes that if an individual perceives a situation or stimulus as threatening or noxious, health may be jeopardized. Writers such as Thomas (221) and Cooley (222) relate these individual perceptual differences to position in the social structure and the experiences that result.

As a result of viewing people as cognitive creatures, a sociological approach sees human needs as shaping perceptions of situations and therefore as an important factor in understanding stress. Needs are learned and sustained through social encounters. Those who are unable to meet demands involving important needs experience stress. Needs differ according to position in the social structure, prior learning experiences, and the conception of the self, which may have been learned as positive and capable or as negative and therefore powerless or helpless.

Class-based distinctions also should be drawn in relation to needs. Resulting differences in income and wealth, education, and life opportunities are evident between lower, middle, and upper classes (see 226). These form the basis for distinctive sets of needs that require fulfillment. For example, the increased morbidity of lower classes means that increased pressure is placed on them to seek solutions to their health problems, differing from those who experience fewer problems.

Those who have developed a sense of helplessness due to their position in the social structure are least likely to satisfy important needs and therefore are most prone to stress and ill-health. Otto (18) has addressed the role of helplessness and a lack of control, treating them as major contributors to stress and ill-health (see 227).

Thus individuals' perceptions are based on their experiences in the social structure. Those with working-class experiences can be expected to assign different meanings to situations than those with middle- or upper-class orientations. Socialization, past experience, and position in the social structure all provide certain opportunities and resources. The degree to which people perceive situations as threatening depends on the degree to which they are not able to meet important needs or do not have sufficient resources to deal with demands.

Sociological Studies and Stress Theory

A group of sociological studies emphasizes the relationship between sociological concepts and other disciplinary approaches. Marianne Frankenhaeuser (10) has developed a psychobiological approach to stress as a way of integrating a number of different approaches (including sociology, psychology, and physiology) to the study of the stress response. She has linked the physiological stress response in individuals to the context of their broader experience (see 108, 109, 137, 228).

Frankenhaeuser argues that "with the development of biochemical techniques that permit the determination of small amounts of various hormones in blood and in urine, psychoneuroendocrinology has come to play an increasingly important role in stress research" and that cognitive appraisal affects neuroendocrine responses "and the emotional impact of the stimuli rather than their objective characteristics" (10, p. 216). The physiological stress response is determined in important ways by values and attitudes.

Frankenhaeuser found that when effort was expended in arithmetic activities under extended noise conditions, adrenaline, noradrenaline, and cortisol levels and heart rate increased. Frankenhaeuser and Lundberg (229) found that, under conditions that induced lower aspirations, the initial low noise had lasting effects on the mode of adjustment. Performance decreased as noise increased and physiological indices did not increase. Frankenhaeuser argues that initial adjustments may be made that reduce the ability to deal with threats. She maintains that this conclusion is based on evidence from a number of sources and further argues that if catecholamine "secretion is prolonged, damage to various organs and organ systems may occur" (10, p. 219). Frankenhaeuser does not sufficiently emphasize the broader sociocultural context within which these individual personal responses take place, even though she has shown the need for developing such links. However, she has added an important understanding to how links between physiological and psychological response and broader social experience can be made.

A series of sociological studies that have contributed significantly to our understanding of stress show that stress is a process and not a series of unconnected relationships: stress is a social process in which groups who experience stress have basic needs that are not met. These groups are disadvantaged due to their relationship to the dominant mode of production in society and consequently due to their position in the social structure.

More recent sociological approaches have pointed to a consideration of many appropriate contextual factors in the analysis and understanding of stress. Pearlin and coworkers (178) have made a significant contribution in presenting a model that locates stress along with other variables as part of a process leading to ill-health. These authors propose that (178, p. 342):

> life events can lead to negative changes in people's roles, changes whose persistence wears away desired elements of self-concept, and that through this set of linkages stress is aroused. Coping and social supports, for their part, can intervene at different points along this process, thereby mediating the outcomes.

Their model accounts for a series of different influences in stress and treats it as a continuing process. Pearlin and colleagues developed a model that included the personal resources of self-esteem and mastery as well as coping resources such as social support to explain stress and ill-health outcomes. These are important interactive concepts in explaining how individuals respond to stressors. However, the model does not account for the influence of social structural factors. As Otto argues (18, p. 9):

> it needs to be placed in a broader sociological context, so that the power structures which produce or affect stressful life and work experiences, and the interest groups which promote system-maintaining coping modes where there is need for change, can be more fully examined.

There are many different causes of occupational stress. They can best be seen as arising out of interactions between individuals with inner needs and conditions in the external environment. When there is a lack of fit between the demands and resources in the environment and the opportunity for people to satisfy their needs, stress occurs.

Stress is most often referred to by researchers as a negative response, or distress. Mechanic argues that "the term 'stress' has been used to refer to emotional tensions—anxiety, fear, depression, general discomfort—either reported or observed, from which it is inferred that the individual is exposed to some stress situation" (63, p. 229). Many stress researchers also conceive of stress as a mismatch between the demands in the environment and the person (e.g., 230, 231). Cox (232) argues that discomfort or stress occurs when individuals recognize they are unable to deal effectively with demands or other conditions in the environment, and that an individual's perception of a lack of fit and the problems this causes is a precondition for stress.

According to Mechanic, "stress characterises a discrepancy between the demands impinging on a person—whether these demands be external or internal, whether challenges or goals—and the individual's potential or actual responses to these demands" (63, p. 292). Stress is a complex process and involves many aspects of interaction between the person and the environment. Mechanic argues further that (63, p. 292):

> when discrepancies develop or are anticipated they are associated with physiological changes, feelings of discomfort and concern. The extent of physiological change and feelings of discomfort will depend on the importance of the situation or the extent of motivation, or the degree of discrepancy of failure anticipated or experienced, and on genetic and physiological factors.

He focuses on both individual capacity to meet external demands and characteristics of the social structure by addressing the various aspects of adaptation to external demands. If adaptation is successful, ill-health outcomes and other negative consequences of stressors will be mitigated. As Otto (18) argues, this not only can create changes in the individual's abilities and capacities, but it can change the extent of the strength and intensity of external demands.

Mechanic argues that "successful personal adaptation has at least three components at the individual level." First, the individual needs the capacities and skills to deal with demands (or coping capacities). These skills "involve the ability . . . to influence and control the demands to which one will be exposed and at what pace." Second, there needs to be sufficient motivation to meet demands: "individuals can escape anxiety and discomfort by lowering their motivation and aspirations . . . as motivation increases, the consequences of failing to achieve mastery also increase," increasing anxiety. Finally, a state of psychological equilibrium needs to be established in order to direct skills and energies (defenses). Each of these aspects of adaptation to external demand is based on the ability of the social structure to provide adequate resources to meet external demands. Mechanic argues that "this fit between the social structure and environmental demands is probably the major determinant of successful social adaptation" (all quotations from 68, p. 33). The ability to exert control over external situations depends in part on individual capacity, but in a capitalist society it is based on the resources and constraints provided by location in the class structure.

Mechanic is one of the few stress theorists to clearly place the individual in the context of a social structure and institutional frameworks by looking at the mismatch between individual needs and the environment. He draws attention to the capacity of social structure to provide adequate preparatory socialization through the family, schools, and other agencies. He also identifies the need for social structure to provide incentive systems that channel motivation and to establish and maintain beneficial social support networks. Eyer (233) also argues that the causes of ill-health can be understood by examining the inability of social structure to prepare people adequately to meet environmental demands, and that the causes of ill-health in modern capitalist society are social and structural.

In order to understand the complexities of the stress response, sociopsychological approaches need to be considered alongside the earlier, purely physiological approaches. The social and cultural context determines first an individual's adequacy to meet demands, then the individual's perception of the importance of meeting those demands, and finally the type of response. It is a vital element in the stress response.

Rosemarie Otto, an Australian sociologist, has made a valuable contribution to stress research, first in investigating occupational differences in stress and ill-health (which few studies have done (see 134)) and later in extensively examining stress among school teachers. She developed links between the physiological response in stress and the broader social structure.

Otto (18) has presented a comprehensive model that accounts for sociopsychological aspects of the interaction of the person with the environment and shows the corresponding responses. She argues that "stress is described as an inner state of a person, which involves both an experience of a tension-producing situation as well as a physiological response . . . stress arises in the process of interaction between person and environment," and she describes the environment as including "demands or constraints imposed on people, the potential 'stresses' or stress-producing conditions people encounter" (18, p. 34). Otto argues further that how much stress a person experiences "also depends on the needs, expectations and capacities (the internal demands and constraints) which he or she has bought to the situation and which will affect his or her perception and assessment" (18, pp. 35–36).

In her model of stress, Otto argues that any imbalance between demands and external and internal resources will create stress. Stress refers to "a state of unpleasant emotional tension" (18, p. 36) or distress. It also involves physiological reactions that are required when dealing with demands or threats. Ill-health results if overpowering situations place undue demands on coping responses and the body reaches a stage of exhaustion. Otto further suggests that whether ill-health outcomes will result depends on a number of external factors including the extent of stressors, availability of support and other resources, and power to exert influence. Certain internal factors, including learned coping skills, knowledge of resources, past positive experience, and perceived powerlessness, will also mediate the effects of stress experienced.

Otto presents a processual model and interpretation that link social structure to perception and ultimately physiological response. However, there are two major problems with her analysis of stress. First, her analysis of the experience of a lack of control and lack of power at work is not placed in a sufficiently comprehensive framework to explain the causes of differences in control at work—a problem common to most studies that focus on the effects of a lack of control at work, and a problem of control in capitalist societies. Second, she describes internal resources and constraints as a response to social structure, an important point also made by Mechanic. However, unlike Mechanic she does not clearly identify the sociopsychological processes identified by interactionist sociologists: these explain more fully just how these needs and resources are established. However, in linking social structure to physiological reaction Otto's model goes beyond many other models of stress. Nevertheless, it is possible to further expand the context and nature of the processes through which social structure produces and sustains occupational stress.

STRESS, ILL-HEALTH, AND
SOCIAL STRUCTURE

A number of studies have demonstrated the effects of socioeconomic and class position on ill-health (e.g., 234, 235). They have shown that those with the least resources are most likely to die at an earlier age and are more likely to develop chronic and acute illness that those in higher class positions. Occupation has been shown to be a major predictor of health status, as well as income, education, and unemployment. Relationship to the means of production explains a great deal about health status.

Not many studies of occupational stress consider the relationship to the means of production—and therefore class—as a way of explaining differences in stress and ill-health. Doyal (236), Navarro (6), and Schatzkin (8) outline the class basis of medical health care. It is shaped and maintained by differences in relationship to the means of production and the production process and serves the interests of those in control. It is maintained by an ideology of dominance.

I will examine here the nature of work and experience of occupational stress and ill-health for different occupational groups. This requires peeling back some of the layers upon which a dominant managerial ideology is based in a western capitalist economy, particularly Australia. The exciting and somewhat daunting task is to look at some of the accepted explanations and justifications of the current organization of work and work relations within the context of a political economy and its consequent outcome of social relations. Management has tended to treat occupational health as the responsibility of the individual worker and often blames workers. If a worker gets sick, he or she must be the wrong person for the job. Illness is often referred to as the fault of the individual worker or as an excuse not to work. Based on some compelling sociological literature, occupational health is a direct result of the nature of production relations and the organization and management of work processes. These arise out of the nature and workings of the political economy.

THE LABOR PROCESS

There are several aims in reviewing the labor process here and in Chapter 4. The first is to describe the process; the second is to establish the ways in which it affects the various levels of work and work experience; the third is to look at the specific ways in which stress is a result of the workings of the labor process at these levels. In Chapter 5 I review the role of the state: how it contributes to a labor process conducive to stress for many workers, and its role in compensating for stress and ill-health.

The labor process approach ostensibly began with the work of Harry Braverman (2) when he revived some labor process concepts from Marx's writings. The labor process is a system of economic relationships, a framework in which social and economic activity takes place. Its relevance for occupational stress is that it describes the broader extent of the nature of work relationships within which work is carried out in capitalist societies. It shows clearly the relationships of subordination, domination, and control that exist between owners and managers, and those who have their labor to sell. For Marx, the defining point of capitalist society was the relationship between capital and labor.

The Labor Process as Seen by Marx

According to Marx, the capitalist mode of production involves the production of value, of surplus value, and of capital, and the reproduction of conflicting relations between the classes. Marx identified two distinct elements of the capitalist labor process (237, pp. 291–292):

> the labour process, when it is the process by which the capitalist consumes labour power, exhibits two characteristic phenomena. First, the worker works under the control of the capitalist to whom his labour belongs. . . . Secondly, the product is the property of the capitalist—not that of the worker, its immediate producer.

The capitalist owns labor power through purchasing it with the intent of creating profitability. This according to Marx, translates into two objectives of the capitalist (237, p. 293):

> in the first place he wants to produce a use-value which has exchange-value, ie. an article designed to be sold, a commodity; and second he wants to produce a commodity greater in value than the sum of the values of the commodities used to produce it, namely the means of production and the labour power he purchased with his good money on the open market. His aim is to produce not only a use-value, but a commodity; not only use-value, but value; and not just value, but also surplus value.

He identifies issues of ownership and control of the labor process as central to an understanding of capitalist society. According to Marx there are three elements of the labor process: "(1) purposeful activity, that is work itself, (2) the object on which that work is performed, and (3) the instruments of that work" (237, p. 284). The latter two elements represent the technological structure of the workplace.

In the capitalist labor process, owners or managers own and thereby control labor with the intention of producing profitability. Thompson argues that, according to Marx, in the early days of industrialization "the goal of capital become the subordination of labour on its own terms" (238, p. 41).

Labor process approaches to the organization of work have been redeveloped from Marx's earlier writings. Despite almost two decades of interest in its constitution and effects, relatively few sociologists have paid attention to labor process analysis. As Burawoy notes "the study of changes in the labour process is one of the most neglected areas in industrial sociology" (239, p. 190).

Braverman and the Labor Process

Following Marx there have been many writers on the capitalist system but little analysis of the labor process until the early 1970s with the work of Harry Braverman (2), which saw the revitalization of interest in the labor process approach. Braverman's work on the labor process and the subsequent development of labor process theory is important because his work led away from some of the earlier sociological studies which concentrated on issues related to industrial psychology of the workplace. Braverman's account of the labor process helped reorient sociological studies toward a more coherent structural analysis of work in modern capitalist society.

The study reported in the later chapters of this book involves an evaluation of management control, its strategies and consequences for occupational stress and ill-health. A labor process analysis helps to identify the consequences of work design and the system of management, as well as the structure within which capital and labor play out their conflict. Labor process theory helps in identifying the context and causes of occupational stress.

Braverman's work was concerned with the historical development of the labor process. In some senses Braverman was simply restating Marx's perspective. Thompson (238) rightly notes that the two trends occurring since Marx's time of writing formed a substantial shift for Braverman in an analysis of the labor process. These two trends were: first, greater opportunities for tightened management control through the spread of Taylorist principles of work organization and management; and second, the new developments in technology and science that led to widespread deskilling. These two elements of increased management control became focal points for Braverman's analysis of the labor process.

Carter (240) argues that the Marxist two-class model does not capture recent workplace change in an appropriate way. In particular, it does not address global restructuring of class and employment relations. He calls for a reintegration of class relations through Marxist and labor process theory.

Following the work of Marx, Braverman identifies a key issue to understanding the capitalist labor process as the conflict that arises over surplus value, its generation, and appropriation. As a result, Willis argues for Braverman that (241, p. 77):

> the central dynamic of the labour process is the realisation of labour power—
> the transformation of labour power potential into actual labour. For the work-
> ing class, who have only their labour power to sell in order to make a
> likelihood, the task is to find an employer who will engage that labour power
> potential or capacity to work, be it physical strength, skill or knowledge, in
> return for a wage or salary. For the employer, the task is to transform the
> labour power potential into actual labour in order to provide wages and
> make a profit.

Braverman argues that the role of management in this process is to gain increasing control over the process of production: to organize production for the meaningful fulfillment of profitability goals and to ensure an effective utilization of labor power by exercising control over and offering some security to employed labor. He describes the labor process in capitalist society as a problem of control (2, p. 58):

> [It] becomes essential for the capitalist that control over the labour process
> pass from the hands of the worker into his own. This transition presents itself
> in history as the progressive alienation of the process of production from the
> worker; to the capitalist, it presents itself as the problem of management.

Braverman addresses the two trends in management practice and technological development since Marx's work: the science of management epitomized in the work of Frederick Winslow Taylor (3) and the deskilling caused by scientific management and technological development in industry.

Braverman argues that Taylorism, based on the work that exemplified forms of management control, is a pivotal expression of the ideology of management to control the labor process. However, he has several critics of his treatment of Taylor's ideas—including Littler and Salaman (242), Burawoy (243), and Hill (244).

The second trend identified by Braverman was the effects of labor process changes on deskilling. According to Braverman, specialization and fragmentation of work occurred as a result of Taylor's scientific principles and the introduction of new technologies (especially mass production) designed to increase management control, and these have led to a deskilling of labor. He argues that as a science, management had its roots in Taylorism which is designed to reduce skill levels on the job and at the same time to give employers greater control.

Technological innovation, which largely has allowed the fragmentation of tasks and the division of labor, is an outcome of the changing nature of capitalism, with individual entrepreneurs attempting to refine their attempts at capital accumulation. In its drive for efficiency it has led to a reduction in skill levels and to greater control by management over the workforce.

Thompson defines three elements of deskilling: first, "the replacement of skilled workers by mechanised processes or machines"; second, divided jobs with specialist workers performing the remaining skilled work; and third, the fragmented unskilled and semi-skilled jobs (238, p. 91). He argues that skill in the labor process is largely based on knowledge, a unity of conception and execution, and control by workers.

According to Braverman, deskilling resulted largely from the separation of the thinking and planning from the execution of the job. This occurred through Taylorism and was a consequence of extreme forms of the division of labor. Technology also formed the basis for separating the conception and execution of work. As technological development increased and was backed by an ideology of scientific management, skill levels were reduced in the workforce and labor was also deployed to other areas of work. This was a means of increasing control over the workforce.

Again there are several critics of Braverman's approach to deskilling (e.g., 238, 244). However, his was one of the first recent attempts to systematically identify consequent skill loss among workers based on management's increased control of the labor process. At times he neglects some important processes such as worker resistance and the need for management in some industries to gain some control through compliance, not just coercive power. But he has had a marked effect on industrial sociology in refocusing issues on the labor process, especially within a Marxist analysis. As Willis points out Braverman's (241, p. 5):

> study served to reorient and reintegrate the sociological study of work towards the more traditional nineteenth century concerns of political economy. Braverman's work has stimulated a great deal of research and analysis and will be judged one of the most important books published in the social sciences in the second half of the twentieth century.

While Braverman's work has been rightly heralded as a substantial contribution to our understanding of the labor process debate, his approach has sparked off critical debate. It is not without flaws.

An understanding of labor process organization is important to understanding the social production of conditions of stress. Coburn argues from a Marxist position that as work is forced, need satisfaction is related only to satisfaction outside work: "in the production process, the needs of workers are irrelevant" and "sufficient evidence exists to relate ill-health to the effects of alienating work on the lack of need satisfaction" (4, pp. 42, 57). Schwalbe and Staples (7) show that those in lower occupational groups with insufficient control at work and too much job routine are more prone to stress, poor self-esteem, and ill-health than those in higher occupational groups. A number of other studies have used the framework of structural relationships to explain stress (e.g., 4–6, 8, 130). These studies show that stress is a social process, an outcome of the social relations of production. In most western capitalist countries, stress and ill-health result from a lack of control over the capitalist labor process. Navarro (245) notes that most Americans now believe that government is run to serve the interests of a few, and that most experience a high level of alienation. The role of government in promoting the interests of a minority with the consequence that many suffer must also be questioned: ill-health is one consequence of this level of alienation. It is only by viewing stress as a broader sociopsychological process that we can see its causes more clearly.

CONCLUSION

Stress requires a broader understanding of social processes. In order to understand stress and ill-health in most western countries, we need to understand the social and structural processes of advanced capitalism. Stress and ill-health can be seen as a necessary cost of an economy seeking to maximize corporate growth and productivity at the expense of fulfilling important needs of workers.

In earlier chapters I have identified the biophysiological and psychological approaches to stress. A purely physiological approach is furthest from adequately explaining the conditions that lead to occupational stress and ill-health, while psychological approaches have several limitations in largely ignoring the sociopolitical context of stress. A sociological approach contributes a great deal to our understanding of the social, cultural, and political framework within which stress and ill-health take place. It also much more fully develops the idea of a total process of which stress and ill-health at work are the outcome.

Stress is the result of negative work conditions arising from management's control over the labor process. Braverman's work began a reconsideration of control over the labor process and a resurgence of interest in work organization outcomes that result from class conflict over labor process control. His work led away from a preoccupation in many sociological studies with an industrial psychology approach. It is an important precursor to understanding the social process of the production of stress and ill-health: they arise largely out of management control of the labor process.

An understanding of these elements of management control helps to explain why occupational stress occurs differently for different occupational groups. Blue-collar workers, especially the semi-skilled and unskilled, are at the point where management exerts a substantial control over the production process, and they also have the least influence over organizational structure and the policy that affects them. This is less so for white-collar workers. They tend to identify with the same class groups as managers and identify more closely with managerial goals: they are the potential source of the management ranks. Their control is affected by technology, but not to the same degree as for blue-collar workers. Managers usually enjoy the greatest control, but this is not always the case for lower and middle management. They often fight to resist reductions in their control by senior management policy initiatives.

CHAPTER 4

Stress, Management Control, and the Role of Alienation

This chapter has a number of aims. The first is to evaluate research that is based on Braverman's labor process analysis. I consider management control in relation to technological change and in relation to more recent theories on management's role in strategic planning. There are certain deficiencies with Braverman's approach. Later writers on the labor process have refined the concept and the utility of its operation quite a way beyond Braverman. The second aim is to look at the role of the concept of alienation in examining occupational stress and ill-health. Much of the literature on stress does not adequately explain its causes or incidence. Alienation is a concept that can better explain the effects of being unable to fulfill important needs at work. It can link the effects of management control of the labor process to individual experience. Third, I review the literature on the sociopsychological concepts of low self-esteem and low sense of mastery. These states are outcomes of management's control over the labor process and add to our body of knowledge on the occurrence of stress at work.

THE CAPITALIST LABOR PROCESS— AFTER BRAVERMAN

An understanding of the causes of stress and ill-health needs to be based on an analysis of the mechanisms of control extended over the labor process by management. The processes through which an advanced capitalist society maintains management's control can give insights into how stress may be caused and sustained. The development of capitalism depends on the ability of firms to compete with one another to accumulate profit and to reinvest. International competition among firms and among industries has increased dramatically during the last two decades, largely due to the overwhelming development of the international marketplace.

Greenlund and Elling (246) suggest that the differing elements of the U.S. production system in the world capitalist economy affect the outcomes of workers' health and safety. They argue that workers in the global sector are more at risk of illness and injury.

A general rhetoric of government and industry leaders has focused on the need for employees to work harder and be more productive in order to speed economic recovery. Employees generally acquiesce to the demands of employers in times of growing unemployment and economic insecurity. Even strong union opposition to increased management determination over the goals of labor has not reversed this trend. As Elling (216) has pointed out, some of the powers gained by unions have been eroded up to the mid-1980s. Capital has managed to gain greater control to ensure that its goals are met at the expense of diminishing needs satisfaction for employees. In improving their position through calling for a consensus between workers and management, corporations influence the relative relationships between workers and the owners of capital. Littler and Salaman argue that (242, p. 34):

> the corporations' efforts to increase profitability . . . may directly involve interventions in the class relations which exist between employers and employees. . . . Despite rhetoric about efficiency and productivity and use of such neutral terms as modernisation and rationalisation, they inevitably and centrally involve attempts to re-order class relations in terms which increase the relative advantage to capital through manipulating the cost of a unit of labour, the relative size of profit.

The ability to accumulate profits and to reinvest has also grown, as evidenced by more than two decades of mergers and takeover activities. This is especially evident in larger corporations. According to Littler and Salaman (242), monopolization has occurred in the western enterprise, with large proportions of the workforce working for very large companies. Also control, ownership, and decision-making are handled by experts, thereby separating the functions of capital from control.

The problem of control over work is central to a labor process analysis. It enables a more fruitful analysis of the ways in which control operates in the workplace. This in turn can lead to evaluation of control at work as an important aspect of work-related stress and poor health outcomes.

Problems with Braverman's Analysis of Control

While Braverman was primarily concerned with control, his conceptualization has a narrow focus. A major problem with his analysis of managerial control over labor is that it deals principally with control of the production process: managerial control can be less obvious. Tanner and colleagues (247) argue that some of the labor process precepts are not substantiated by evidence. These authors are critical of the emphasis on point-of-production activities as the sources of consciousness in capitalist societies.

There have been many analyses of control that apply to management control over work processes. Workers' agendas can be important in determining autonomous behavior. Hodson (248) suggests that workers use resistance, creative autonomous efforts, and compliance to structure their own work activities. A struggle for control is at times against coworkers as much as against management.

Sharma (249) reviewed a number of different approaches to self-management, participation, and control. He also identified the need to instill a culture of participation for human resource development to be effective. Peer supervision in quality circles and the increased power of management information systems has meant that "just in time" and total quality control are efficient forms of discipline and self-control with a minimum of supervision (250).

There are examples of improved work practice not being brought about by improving quality of worklife, however. Russell (251) argues, based on a study of a Canadian potash mine-mill, that downsizing and job expansion have brought about more responsibility and less dilution of jobs.

Boswell (252) suggests that a theory of transactional cost analysis is useful for understanding increases in worker control. Under transactional cost analysis there are three conditions for increased worker control in labor market management: first, work is independent and employees have credentials in the organization; second, employees have credentials and information is available in workers' favor; third, asset specificity accompanies worker control.

Management Control

A number of writers have reviewed the nature of management control using labor process theory. Story (253) suggests that labor process writers have been influential in explaining managerial control over work.

Kivinens (254) argues for a revision of Marxism on the middle class, with different types of mental labor as the basis of approaches and resources of power in production politics. He argues for a distinction between marginal and core groups in the middle classes, with core groups quite distinct from the working class and marginal groups containing many working-class elements.

Littler and Salaman (162) point out that while Braverman recognizes the difference between labor power and labor (labor power is exchanged in the market while labor is exchanged in the production process), he does not bring it into his analysis. This has implications for the way that he treats the problem of control. That is, the subordination and domination of labor should be viewed as real possession instituted through a mechanism of control. Control is a critical concept and Braverman treats control in a monolithic way, centering on the actual production process.

Littler and Salaman expand on Braverman's concept of control in some important ways. They argue that there are several levels at which the nature of management's control of labor needs to be addressed, over and above the production process (162, p. 260):

> It is only through an analysis of the ways in which all levels of employees participate and play some part (foremen, supervisors) in modifying and interpreting, formal organisational controls and objectives that we can understand the relationship between senior organisational members' objectives and decisions and actual organisational outcomes. Similarly, it is only through the processes of adjustment of the formal specifications that we can understand the development of what are, from the shop floor more congenial arrangements, which can then be treated as, to some extent, stable patterns of accommodation.

They add that "a more useful theory of labour processes cannot be restricted to the specification of work activities at the point of production itself but must take account of the control implications of decisions taken elsewhere in the organisation, and indeed outside it" (162, p. 260).

Looking at control other than control by management at the point of production is an important departure from Braverman's approach. Other modes of control, both technical and bureaucratic, exist throughout the structure of organizational activities. Management works out the arrangement and coordination of work processes and establishes the best means for controlling labor. This level of control can extend to activities and decisions beyond the immediate firm: these can involve investment, merger, and other decisions that may have implications for the control of labor. Finally, management decisions affecting market growth and competitiveness can affect short- and long-term employment opportunities: these activities of management can therefore exert a marked control over labor opportunities to work. In fact, very few studies of occupational stress have extended the concept of control beyond the production process. The focus of much occupational health has been on the effects of production-line control rather than broader effects of a lack of control.

Direct and coercive control by management is not the only way in which the labor process is influenced. Management often seeks compliance from labor to work toward common goals (255). Labor also acquiesces to managerial control. Unions often strike bargains with employers in which they agree to accept a certain level of domination if some important workers' needs are met. Berlinguer and coworkers (256) point to ethical problems in relation to work and health. The differences in power, control, and ethical considerations between workers and employers highlight conflicting interests in the health of workers, the environment, and the general population. Burawoy (1) and Littler and Salaman (162, 242) argue that as capitalism develops, the labor process becomes the tool primarily of capital and workers lose substantial elements of control. Burawoy argues further that through the separation of conception and execution, labor becomes further alienated from the labor process.

It would be inappropriate not to address one of the principal vehicles through which management has managed to widen the gap between conception and execution of work in the relative functions of management and worker. The wave of new technological developments in industry has widened the rift between the work experience and relative working conditions of occupational groups such as managers and white-collar workers, and blue-collar workers. This indicates some of the less direct and less obvious ways through which management enlarges its area of control. These increases in control have important effects on the experience of occupational stress and ill-health.

THE ROLE OF TECHNOLOGY

Some of the effects of technological development (machine pacing, VDU systems) on stress and ill-health were reviewed in Chapter 2. However, the broader effects of technological development, which increase the means of control by management over the work experience, have received scant attention in occupational stress and health research.

The scientific-technical revolution following Taylor saw increases in productivity resulting from increased mechanization and technological development. Technological development has occurred as an attempt to meet the demands of increased productivity through increased production. Marx (237) said that control becomes secondary with the introduction of technological innovation. However, the cries for increased productivity through technological development have a certain class basis of control to them, with one class attempting to exercise control over the other with the obscuring ideology that a technological necessity is embodied in this.

In most cases, it is in the interests only of capital to secure greater control through increased mechanization. Workers lose their jobs with increasing mechanization; they have less control over the whole job because of the accompanying division of labor between workers' groups and between management and workers; and generally, society suffers from the impact of technological development through increasing pollution. Braverman (2) and later Littler and Salaman (162, 242) and Child (161) were quite clear in identifying the ideological basis of management's increased control through the rhetoric of a "technological imperative." This was often expressed as fulfilling a need for competitiveness, growth, and survival: the needs of increased technological development, by necessity, must be met. Child (161) argues that this strengthens management's claim for control: the important factor is the nature of the policy that develops and management's claims for its rationality and need.

There are two problems with Braverman's analysis of management control of the labor process. First, he neglects some of the more subtle means through which control is gained, and second, as noted earlier, he concentrates on control at the point of production instead of operationalizing a wider idea of control. Child argues that management exercises control over the labor process through the development and implementation of strategy and of strategic decision-making. These strategies are stated in relation to achieving more general capitalist objectives. According to Child, "in capitalist economies corporate management strategies . . . reflect a consciousness of certain general objectives which are the normal condition of organisational survival," and these are generally described by senior managers "in terms of profitable growth . . . [they] are not necessarily formulated with the management of labour and structuring of jobs specifically in mind" (161, p. 232).

For instance, new technology can be introduced to strengthen market position or, as Burawoy (239) points out, to force smaller capitalists out of the market. Its implementation, however, can result in a loss of control, discretion, and skill use in the workplace for workers. The process of management gaining control and workers having to make substantial concessions is not as direct and straight-forward as early labor process writers conveyed. Braverman looked specifically at direct and coercive control by management, while Edwards focused on compliance. For Child, policies unspecific toward the labor process have relevance for it—that is, a policy of market development may in effect lead to specific reductions in control for lower level employees.

The commonly stated purpose of managerial strategy, according to Child, is profitable growth in the context of product and labor market concerns, and development of technological know-how. Objectives such as improving control and efficiency and reducing costs become operationalized. I would argue that for the various levels of management this can affect the reorganization of structure and work processes such as workflow design, the structure of reporting relation-ships, and workload requirements. The implementation of these objectives can have specific effects on control and skill. This can occur especially for lower level managers and supervisors and for operative workers.

While Braverman argues that control over the labor process is a fundamental objective of management, Child argues that managerial control of the labor process is implied as part of a more fundamental capitalist objective of management. I have suggested that in order to understand the broader issues of control in the workplace, a wider theoretical position than that adopted by Braverman and some other labor process writers is appropriate. Issues of control are related to issues of class membership. Managers perform a role similar to that of past owners in that they seek to achieve capitalist objectives related to the accumulation of surplus value; in doing so they tend to espouse a managerial hegemony similar to that of owners. This hegemony states that it is in the interests of workers to place the needs of the organization before their own, and that certain of their needs are likely to be fulfilled by working toward the goals of management. Coupled with this ideology is the growing belief among workers that management's need and right to control is essential for industrial success.

A broader analysis of the ways in which control is established and maintained can show that it should be considered as a vital element in the occurrence of noxious work factors that lead to stress and ill-health at work. A thorough approach to occupational stress needs to be located in a structural framework based on an analysis of conflict over control of the labor process.

This study of occupational stress treats the effects of the labor process on work at three levels. The first effect is in restricting job latitude and the opportunity to exercise control over the job. This affects latitude over point-of-production activities. Routine work is a result of job fragmentation. When an employee sees little of the total product or service, there is less opportunity for fruitful decision-making about the job. This also creates a dependence on management control. When no new skills are developed and previously learned skills are not used, a debilitating underload can result. According to Braverman this can stem from deskilling. When there is a perceived lack of control, management and work process requirements dominate.

The second effect manifests through the bureaucratic structure that management establishes for decision-making and the control of work process. Greater bureaucratic structures arise as management attempts to gain greater control over organizational processes, particularly the decision-making and informational processes. This type of management control is less evident to workers, but affects their work experiences. Management makes decisions and creates work structures and processes in a number of areas that affect workers' experience. These include supervisor and subordinate relationships, which can result in role ambiguity and conflict, and management of the structure of workgroup formation and interaction around the work process, which can influence the quality of work relations. Other areas are job security and career paths or promotion opportunities offered; these affect rewards and other benefits for workers. Decisions to increase control are usually taken by management to improve competitiveness in the marketplace, and most often at the expense of workers' needs.

The third effect of management control is reflected in the "management prerogative": it also creates a climate of attitudes and values in the organization. Management's position of control can be expressed by negative attitudes toward workers, poor feedback, and a general lack of appreciation. It reflects the workers' position of powerlessness under a capitalist labor process. This lack of control by workers is evidenced in the relationships between workers and management and it constitutes the day-to-day climate under which work is carried out. Finally, workload requirements also stem from management decision-making and policy creation to improve or maintain competitiveness in the marketplace. Frederick Taylor (3) specified methods for management to use in order to gain maximum output for the smallest labor costs. Many similar ideas are still current in management practice and stem from treating labor as a malleable commodity. Workload requirements seldom stem from meeting labor needs or the mutual needs of management and labor.

A lack of control facilitates the development of noxious work experience. Each of the factors outlined above results in important needs not being met by workers, which in turn is likely to lead those workers to experience stress and possibly ill-health. Attempts to deal with the symptoms of a lack of control will not address the real causes of stress.

LABOR PROCESS APPROACHES TO THE STUDY OF OCCUPATIONAL STRESS

Some studies that contribute to an individualistic notion of the causes of stress focus on characteristics of workers and their propensity toward stress. The broader stream of studies which have enhanced our understanding of the work conditions that create stress are those focusing on the nature of work and working conditions. Yet these studies are limited in explaining the nature of the causes of stress at work and lead only to piecemeal solutions. Few studies relate the nature of work, working conditions, and management policies and practices to the labor process. These few studies view stress as emanating from working conditions that result from competition for control over the labor process.

A relatively small body of labor process writers have addressed the issue of the broader structural context of the experience of occupational stress and ill-health. They argue that management's increasing control of the labor process has led to more alienating conditions for workers with the least influence and control. As a result these workers experience stress and higher levels of occupational ill-health. According to Navarro (6, p. 15):

> changes in the labor process in large-scale industry were not a mere result of the capitalist's work, but, rather, an outcome of the class struggle carried out under the dominance of the capitalist class. In that struggle, the capitalist class aimed at deskilling the laborer and dividing the working class.

Navarro concludes that "in the process of class struggle, the capitalist creates the conditions that enable the expropriation from labor of their labor power and their health" (6, p. 26). Labor process writers elevate the differences in work relations between management and workers to the level of class struggle. They focus on control as an issue in relation to the labor process and to consequent stress and ill-health. Navarro argues that the "tendency towards losing control over one's work is . . . the characteristic of the labor process under capitalism (6, p. 16). As Willis argues, the labor process approach explains work as a "relationship of power for most people in our society; that is it is structured in terms of a relationship of domination and subordination" (241, p. 4).

The structural framework used by labor process writers on stress and ill-health provides an important explanation of the causes of stress. These causes are located in the context of control over the labor process. Few studies, however, have investigated from a labor process approach the alienating effects of a lack of control over structural factors and the effects of management control over the organizational climate on stress and ill-health. These studies ensure that explanations of the experience of stress and ill-health are firmly ensconced in an analysis of class relations and competition for control of the labor process. However, the lack of strong relationships found between routine, fragmented work and a lack of control and stress, ill-health, and well-being demonstrates that other aspects of alienating work also need investigating.

MANAGEMENT CONTROL AND OCCUPATIONAL STRESS

Labor process theory has made important contributions in the fields of the sociology of work, occupational health and safety, and occupational stress. For the purpose of this book, the labor process is dealt with at a number of distinct levels of analysis. Management's control of the labor process at these distinct levels creates greater stress for workers. The first level, the nature of the work itself, is represented by the task. The various features at this level, include the design of the job, the way work is organized, breakup of tasks, level of routine, amount of skill required, and degree of discretion to exercise control over the job.

The second level deals with organizational structure. It incorporates the process used by management, methods employed by management to organize labor for fulfillment of goals, and division of labor between workers, such as demarcation between groups of workers. It also includes workload demand, management control procedures, and the relationships from these. Finally, it includes the climate of attitudes and values. Bureaucratic values can lead to workers not being accepted or appreciated. It is only their labor power that is accepted by management.

Littler and Salaman (162) argue that a level of control and influence beyond the organization or enterprise operates at the macro or societal level. This is the third level of analysis. It involves management, labor, and state relations in determining and influencing conditions of work activity and employment in the broader economy. National employment strategies, labor-management relations at national and state levels, and national marketing strategies are all linked to national economic goals: these are influenced by international competitive activity. This macro level of the labor process influences the planning of levels of business activity: this affects workers' security of employment and future growth and development in industry.

Each of the two micro levels described above is affected by an interaction between management, labor, and the state. At the level of the job, the state intervenes and prescribes certain minimum requirements such as safety precautions. The state also intervenes at the level of organizational structure by ensuring certain minimum conditions of employment such as minimum wages and benefits. Management control at each level of the labor process activity—the job, and management structure and climate of attitudes—can lead to occupational stress for lower occupational groups because they offer least resistance. It is not new to present occupational stress as an outcome of the way in which work is organized; however, few have attempted to look at the role played by the state in providing conditions conducive to both the production and prevention of stress.

ALIENATION, OCCUPATIONAL STRESS, AND ILL-HEALTH

The concept of alienation is important for occupational stress and ill-health as it helps to explain a set of complex relationships. The central feature of labor process arguments on the relationship between work, stress, and ill-health is that work is alienating. Alienation is a difficult concept. Thompson argues that industrial and organizational sociologists have not adequately addressed the problem of alienation at work: "it is hardly surprising . . . that an understanding of alienation has been weak" (238, p. 22).

Marx treated alienation as a necessary and inherent condition of the capitalist mode of production. He wrote (237, p. 716):

> on the one hand, the production process incessantly converts material wealth into capital, into the capitalist's means of enjoyment and his means of valorisation. On the other hand, the worker always leaves the process in the same state as he entered it—a personal source of wealth, but deprived of any means of making the wealth a reality for himself.

Marx argues that the workers' labor has been alienated before entering the production process. Their product is objectified and therefore estranged from them. Their product, he wrote, is also converted into capital, and (237, p. 716):

> the worker himself constantly produces objective wealth, in the form of capital, an alien power that dominates and exploits him; and the capitalist just as constantly produces labour-power, in the form of a subjective source of wealth which is abstract, exists merely in the physical body of the worker, and is separated from its own means of objectification and realisation; in short the capitalist produces the worker as a wage-labourer. This incessant reproduction, in perpetuation of the worker, is the absolute necessary condition for capitalist production.

For Marx the worker was alienated in a number of ways: alienated from the products of labor, from the act of production, from others, and from their very nature. Because of the extent of control of capital over production and the marketplace, workers do not get the opportunity to meet basic inner needs. According to Marx, individuals are separated from their real nature because of the nature of capitalist society. In a managerially controlled labor process, a reciprocal relationship between the individual, nature, and industry is not possible. According to Thompson (238), however, Marx's concept of alienation is not well suited to present industrial society.

The notion of control is important to understanding alienation. Hill (244) is critical of Blauner's (211) work on alienation for the very reason that it does not sufficiently account for workers' lack of control at the point of production. Alienation is a term that aptly describes the condition of those with too little influence and authority: it is intimately tied to the concept of control. The process by which occupational conditions influence stress and ill-health cannot be understood without a concept such as alienation. At the theoretical level, this concept is an important tool because it explains differences in influence over the means of production and in accessing resources to meet important needs. At the operational level, it reflects levels of influence and control for workers. People are not always aware of being alienated, however, and this poses problems in using this concept.

Coburn (4), a labor process writer, argues that alienation results from a lack of control. Other labor process writers (5, 7) also refer to lack of control as a central feature of alienation. Garfield (5) restricts his study of alienation to a lack of control over the task aspects of work. He examines repetitive and fragmented work in which workers have little control over the task, and argues that "evidence suggests that people experience increased stress when they lack control over the consequences of their own activities (as when they receive inadequate feedback). . . . This resembles the condition of alienation" (5, p. 552). Schwalbe and Staples (7) have also looked at the effects of alienating conditions on the task itself (a lack of control and job routine), and found that routine work with little control were determinants of job stress and low self-esteem. These in turn had important effects on ill-health. Coburn also found that where job alienation is treated as a person-job mismatch or as repetitive and monotonous work, it is "associated with lowered psychological and physical well-being" (4, p. 54).

There are many reasons why the level of alienation at work is greater among blue-collar workers than among other occupational groups. Shostack (195) outlines a number of sources of blue-collar worker alienation: for example, blue-collar workers complain of too little autonomy at work, too little challenge, poor compensation, and limited control of their tasks. Alienation, he argues rightly, is not restricted to point-of-production aspects: other features of the general work environment produce alienation. These include too little health and safety protection, too few promotion opportunities, poor security, and limited status from the job.

Thompson criticizes much industrial sociology research for ignoring prior socialization of workers. Family, schooling, and general class experiences develop sets of needs that workers expect to fulfill at work. Thompson (238, p. 22) argues that this research has "ignored prior orientations to work which strongly influence reward expectation and satisfaction experienced" (238, p. 22).

A particular problem in explaining alienation, especially that experienced by lower classes, is that not all workers expect to satisfy needs such as the need to exert influence and control. I argue that this can partly be explained by prior socialization which leads some workers to restrict their expectations. The hegemony of management control also helps explain the experience of alienation. The dominant rhetoric in capitalist society encourages a belief in the right of management of control. Even though workers' inner needs may not be satisfied at work, in due course they may seek to satisfy secondary or alternative needs and may accept that other needs have to be satisfied in other, out-of-work activities. Often they may accept the right of management to control, despite being unable to satisfy some important needs. This is a difficult and complex problem, but not without explanation. During socialization people internalize the values of capitalism and adopt them as their own. Managerial and government rhetoric encourages the belief that only through meeting the needs of capitalism is it possible to achieve personal need fulfillment, even though a lack of evidence of successful need achievement among lower classes suggests this does not happen. As noted in the discussions earlier in this chapter, political discourse is supportive of the idea that the only way to economic recovery and to ensure their own survival is for employees to work toward corporate needs for productivity and profitability increases. Managerial ideologies, however, permeate through the lower classes. So it is a self-perpetuating problem: unless workers have been socialized or resocialized into challenging the legitimacy of dominant structures and ideologies, little consciousness of common class identification or membership develops.

Nonetheless, if alienation is treated as a consequence of a lack of control over the labor process then the concept is useful, despite some problems of prior socialization and reward expectation. It can usefully explain why stress occurs at higher levels for those who do not have sufficient control and resources to meet work demands adequately and to satisfy important inner needs. Those in lower occupational groups that have low levels of need satisfaction for job elements such as control, skill use, task variety, recognition, and satisfying relationships are more likely to experience the highest levels of stress. Because of prior socialization and a belief in the right of management to control, these may not always be expressed as strong needs.

SOCIOLOGICAL STUDIES AND OTHER SOCIOPSYCHOLOGICAL STATES

Relatively little research has been conducted on the interconnections between stressful events and the outcomes of self-esteem, sense of mastery, and stress levels and how these influence one another. Also there is little reported evidence of the roles of coping and social support in affecting self-esteem, sense of mastery, and stress levels.

Stress, however, is not considered as a result of sociocultural forces alone. Two sociopsychological variables, self-esteem and sense of personal mastery, help to explain the experience of stress and associated ill-health at work. Israel and coworkers (101) argue that sociopsychological variables modify the relationship between occupational stress and ill-health. Personal resources such as mastery and self-esteem can be included as important mediating variables in that relationship. These authors state that a consideration of these variables "is important not only to proper scientific specification and understanding of the relationships between work stress and health, but also to applied efforts to reduce work stress or improve health" (101, p. 164).

Self-esteem and the concept of mastery are two sociopsychological dimensions that have been linked, along with stress, to work experience and ill-health outcomes. These concepts are relatively undeveloped in stress research, and sociologists have paid little theoretical attention to their role in the relationship between occupational conditions and ill-health. At least three observations are possible. First, stress researchers have not yet sufficiently examined the joint interactions of these variables in the stress process. Israel and coworkers observe that "rarely are two or more of these studies considered together in a single study (101, p. 164). Second, rarely have self-esteem and mastery been examined as outcomes of work experience, and third, sociologists have paid scant attention to the effects of organization and control of the labor process on self-esteem and personal mastery.

According to Rosenberg (257), the self-concept measure "self-esteem" derives from evaluations made by the self and by others. It forms from individuals' evaluations of their own competence, from the appraisal of others of social performance, and from social comparisons. In a study of work experience and health, Schwalbe and Staples (7) investigated the role of both stress and self-esteem; as outcomes of negative work experience, they are further evaluated for their effects on health outcomes. As a mediator between work experience and health, self-esteem is described as deriving from "positive reflected appraisals, favourable social comparisons, and self perceived competence" (7, p. 588). Rosenberg suggests two sources of self-esteem: reflected appraisals, being those assessments derived from the opinions of others, and those deriving from feelings of competence. Staples and coworkers (176) argue that a sense of competence ensuing from effective performance is an important additional element of self-esteem. Pearlin and Schooler, who employ a measure of self-esteem in their study, observe that (99, p. 5):

> psychological resources are the personality characteristics that people draw upon to help them withstand threats posed by events and objects in the

environment. These resources, residing within the self, can be formidable barriers to the stressful consequences of social strain. Self-esteem refers to the positiveness of one's attitude towards oneself.

Of the three personal resources employed in their study (self-esteem, personal control or mastery, and self-denigration), self-esteem was the second greatest mediator on stress outcomes.

The concept of personal mastery (or belief in an internal locus of control) was refined by Pearlin and Schooler, who argue that mastery is one of a number of resources "that people draw on to help them withstand threats posed by events and objects in the environment," and "concerns the extent to which one regards one's life chances as being under one's own control" (99, p. 5). It is a socio-psychological state developed through evaluating one's ability to meet important needs and to influence external conditions.

Few sociological studies treat these three concepts together. Self-esteem and personal mastery help to explain the experience and effects of stress. Self-esteem explains the state of the self-concept resulting from the success or otherwise of meeting important needs in the face of external demands and limited external resources. Mastery is the belief that one can meet needs under demands made by the environment and can effect changes to the environment. Stress usually results when there is a mismatch between inner needs and external conditions. Pearlin and Schooler place mastery as a mediating "psychological" variable, along with self-esteem and self-denigration, between stress and occupational strain, hence mediating the effects of strain. They further argue that "resources embodied in self attitudes can help blunt the emotional impact of persistent problems" (99, p. 12). In a later study of the stress process, Pearlin and coworkers (178) found mastery to be a directly influencing variable on stressfulness outcomes.

Negative work experience can have a debilitating effect on a person's sense of control in influencing the world. This will in turn make the individual vulnerable to stress outcomes. Pearlin, however, found that mastery had the least impact on occupational strain of the three "psychological" variables. According to Mechanic "almost all stress investigators, irrespective of their orientations, neglect consideration of the relationship between social structure and mastery" (68, p. 32). Self-esteem and mastery or sense of personal control are formed through socialization and experiences prior to the work experience. However, people at work have opportunities to fulfill important needs such as demonstrating competence in tasks (including adequately meeting the demands of the job), exercising discretion and influence over the work environment, and meeting important needs for sociability and security. When these experiences are combined with positive feedback and appraisals, self-esteem and sense of personal mastery are developed positively. While few studies have included both self-esteem and mastery in examining the effects of work experience on stress and ill-health, still fewer have examined them as outcomes of work experience.

Schwalbe and Staples (7) study is one of the few pieces of research, sociological or otherwise, to do this. They treat self-esteem as a consequence of the experience of control and routinization in the workplace. They argue that self-perceived competence is most affected, that "to derive self-esteem from competent performance . . . requires that an individual produce intended effects on the world, perceive him or herself as the agent responsible for those effects, and value the effects perceived," and that "any form of social organization that severely limits autonomy and control, obscures the consequences of action, or makes those consequences difficult to value will deny people opportunities to derive self-esteem from their efficacious action" (176, p. 588). When considering negative structure, with its resulting social relations and negative management attitudes, self-esteem from appraisals from others and from social comparisons will also be markedly affected. Similarly, I argue that the opportunity to exert influence over the work environment to meet needs will affect individuals' sense of personal mastery. In their study of the stress process, Pearlin and coworkers (178) showed that negative work experience can deplete workers' sense of personal mastery. This can occur through insufficient control over the job and work routine, a negative work structure, or a climate of negative authority relationships and attitudes from management.

At work, people who occupy positions of greater influence and opportunity often come from socioeconomic environments offering opportunities for need fulfillment. Opportunities in schooling and upbringing and associated economic resources can produce high expectations and opportunities to demonstrate competence. For such people, the rewards they have received develop in them a strong sense of self-worth (self-esteem) and a strong sense of mastery and personal control. The rewards they receive at work allow them access to resources that are vital for maintaining a positive self-concept. Managers are in positions of influence and control. They have opportunities to meet needs for exercising discretion and influence over the job and work environment, performing meaningful and valued tasks competently, gaining material and other intrinsic rewards, and gaining feedback and positive appraisals on the performance of valued work. Because of this, self-esteem and sense of personal mastery are most likely to be high. Factory workers are expected to bring to work lower expectations of need fulfillment because of their lower level of educational opportunity, fewer rewards and positive experiences during schooling and upbringing, and more limited access to vital social resources during socialization due to a lack of economic resources. They are also expected to have a lower sense of personal control and mastery and a lower sense of self-worth (self-esteem). Because of a lack of opportunity at work to demonstrate competence and to exert influence over the work environment, lower economic (and intrinsic) rewards, and most often poorer feedback and lower appraisals already, low esteem and mastery are further eroded.

In studies of stress and ill-health, sociopsychological resources are seldom linked to wider class experience. Little attention has been paid to the effects of the capitalist labor process on shaping self-esteem and mastery. I will argue that these concepts derive to a large degree from position in the class structure.

The capitalist labor process can prevent lower occupational group members from experiencing positive self-esteem and a sense of mastery at work. Schwalbe and Staples (7) are some of the few sociologists to examine these concepts in relation to the labor process. Control of the labor process by management leads to fewer opportunities for lower level employees to meet needs. They also have fewer opportunities to develop high levels of esteem from their work experience, and are least likely to develop a sense of control and mastery. Schwalbe and Staples argue that "under capitalist relations of production, deriving self-esteem from efficacious action is difficult for those in working class positions precisely because the capitalist labour process denies them autonomy, control and opportunities to demonstrate competence in challenging work" (7, p. 588). Workers often manage other ways, such as games at work, to meet some of these important sociopsychological needs. This has been dealt with by Baldamas (258) and others.

Changes in management control have been accelerated by developments in mechanization and automation. Sociologists such as Braverman (2) have argued that this has led to less reliance on skilled craft groups and therefore a reduced sense of competence stemming from work for some groups, notably blue-collar workers. For example, during the 1910s and 1920s more sophisticated machine processes were introduced into many manufacturing industries in Australia. Many skilled trades workers were replaced by unskilled women and children, cheap forms of labor. Evidence exists to support the notion that those in lower occupational positions where need fulfillments are not achieved have accompanying low levels of self-esteem and well-being (e.g., 7). These groups of workers would not be expected to derive anything but the minimum satisfaction of basic needs. Self-esteem and personal mastery could not be expected to flow from jobs that require little training and the most basic of skills and routine work.

Two main issues arise in reviewing literature on the self concept. First, the self concept has been included in many studies as an important mediator in the sociopsychological process of stress and consequent ill-health. There is little agreement in the literature on the position of such sociopsychological variables as the self concept in the interactive process. In some studies reviewed in Chapter 2, self-concept variables act as mediators between general status and work variables; in others they are placed between work variables and stress; in still others they result from the experience of distress. There is agreement, however, that their positioning studies is quite arbitrary. Second, many different tools have been used in measuring the self concept. Rather than being problematic, the range of measures used adds to our understanding of the functioning of elements of the self concept in the stress and ill-health process.

The first issue is a problem in studies employing the concepts of self-esteem and mastery: the position and role played by these sociopsychological resources at any specific point in the development of stress and ill-health raises questions about the relevance and type of influence. Israel and coworkers (101) as well as Pearlin and coworkers (178) indicate that the position and therefore type of influence on self-concept variables is quite arbitrary. Israel and coworkers argue that there are limitations in distinctly ordering a series of causal relationships. They "recognise that the perceived job stress could be posited as occurring prior to the 'moderating/mediating' variables (i.e., personal resources, participation, influence, interpersonal relationships," and add that "personal resources might be the result of rather than the determinant of these other psychosocial factors" (101, p. 169).

These concepts may best be studied in a model that treats them as having a reciprocal causal relationship. Israel and coworkers (101) examine the role of personal resources (including mastery) as they affect satisfaction with participation at work: this variable is entered in the model immediately following ascribed and achieved status variables. This is well before the outcome of job stress, and before its inclusion in the Pearlin and Schooler (99) and Pearlin and coworkers (178) studies. Pearlin and coworkers (178, p. 351): suggest mastery and self-esteem intervene at various points in the stress and ill-health process (178, p. 351):

> A personality theorist might find reason to regard the self as the initiating force in the stress process. These are certainly accepted approaches. . . . Some of the relationships that we presented as unidirectional are likely to be reciprocal. Obviously, multidirectionality must eventually be taken into account to capture more fully the complexities of the process through the life course.

Schwalbe and Staples treat self-esteem as an outcome of work experience affecting competence and self-appraisal of sense of worth, and suggest that self-esteem "affects health in predictable ways" (7, p. 583). In the same way, sense of mastery can be treated as an outcome of work experience where a sense of control over external events can be affected by control in the workplace. Both Schwalbe and Staples and Pearlin and coworkers identify symptoms (somatic and/or emotional) resulting from these sociopsychological states. In the study presented in this book they are treated as mutual outcomes of work experience and are tested for their effects on symptoms. Directions of the various interactive effects between them are not tested in this study.

The second issue that affects the outcome of any influencing variable and the way that its effect should be interpreted is the problem of differences in the type of measure operationalized. Studies such as that of Israel and coworkers (101) employed a mastery scale composed of only three of the seven items used by

Pearlin. This need not be problematic if similar factor structures emerge; evidence was not available from Pearlin or Israel on factor structure, reliability, or validity. The use of comparative scales is, however, not a great problem. Schwalbe and Staples (7) and Pearlin and coworkers (178) also use quite different measures of self-esteem.

CONCLUSION

In order to understand how advanced capitalist societies maintain and sustain management's control over the labor process, this control needs to be seen as part of management's pursuit of its objectives of growth and capital accumulation. The goal of satisfying the needs of workers diminishes in this process. In the course of increasing profitability, business enterprises can intervene in and reorder class relations: stress and ill-health for workers is a likely result. The approach taken by Braverman (outlined in Chapter 3) identifies management control as a focal point in understanding the structure and quality of work relations. Braverman's analysis, however, has several critics. In particular, his analysis of control focuses on the effects of control on point-of-production activities. This remains a limited focus for labor process changes. The work of Littler and Salaman (162, 242) and Hill (244) identifies wider changes to the administrative hierarchy and effects on conditions of employment as important outcomes of management's increased control over the labor process. Littler and Salaman, for example, identify two areas—bureaucratic control over structure and the employment economy—where management control has particularly important effects on work experience. A number of writers have shown that management can exert increased control through technological development. Part of the hegemony of management control can be seen through the technological imperative: that is, the necessity for introducing systems of production and management that further enhance management's control. Increased stress and ill-health can occur as primacy is placed on the need to achieve management's goals, often at the expense of workers' personal goals. This argument extends the effects of management's control beyond that proposed by Braverman. A second major problem with Braverman's analysis is that he does not account for some of the subtle ways in which management's control is established. Child (161), for instance, also shows how the whole decision-making and policy formulation process of management increases its control, reducing workers' opportunities to fulfill important needs at work. The implications for a broader analysis of control are that it has far-reaching effects on stress and ill-health, beyond its effects on point-of-production activities.

Labor process studies specifically treat stress and ill-health as outcomes of competition for control over the labor process. Stress, they argue, results from processes inherent in a class struggle: in capitalism, losing control over work is a likely condition for workers. In this process, the concept of alienation is important for understanding the ways in which negative work experience causes stress. Coburn (4) and others argue that workers become alienated, that they do not satisfy important needs at work when labor process organization provides insufficient opportunity to exert influence at work. Not all workers are aware of being alienated, however. The operation of a concept such as alienation can be better understood in the context of managerial hegemony. Management is able to maintain dominance by presenting corporate (and its own) goals as being in the interests of all. If alienation is treated as a consequence of a lack of control over the labor process, it can be useful in explaining why stress and ill-health are distributed differently among occupational groups.

The incidence of occupational stress and ill-health can best be understood as an outcome of social structural processes that create alienating working conditions. These result from management's control of the labor process. It is only in this context that physiological responses and psychophysiological and psychological processes can be adequately understood.

CHAPTER 5

Stress as Part of a Social and Political Process

This discussion of the sociopolitical context of stress looks at the role of the state in both contributing to a labor process conducive to stress for many workers and compensating for stress and more broadly for ill-health. It also considers the state's role in establishing and maintaining class differences. Government legislation and policy establish and maintain ideology and create pathways in the capitalist labor process to support and strengthen class differences between owners/managers and workers. The process of production is not merely about economics; it also has political and ideological aspects. I also discuss the findings of Willis (260) and others that important social and political processes establish ill-health conditions as acceptable by insurance agencies, compensation courts, and the medical profession. This is important in explaining how an occupational illness gains support and becomes a publicly accepted and legitimate medical issue. Littler and Salaman (242) support this by arguing that the process of production also includes political apparatuses which reproduce these relations of the labor process through the regulation of struggles.

In a number of western capitalist countries such as Australia, a premium has been placed on work organizations becoming highly productive and competitive. The way in which work organizations are controlled and managed, and the effect this has on people who work in these organizations, is of interest in this chapter. A Marxist perspective shows that the owners and managers of work organizations form a more privileged class. Those who sell their labor are generally from a more working-class orientation and tend to work toward goals that are not of their own making. The types of control used by owners and managers of work organizations and the rhetoric that supports this control are of primary interest in explaining the occurrence of occupational stress and ill-health. In addition, the hegemony underlying their position of dominance is in need of explanation and evaluation. The state, whose role is of immediate concern in this chapter, also plays an important role in fostering a position of dominant managerial goals.

Certain key concepts are crucial to understanding a critical sociological perspective of occupational health and safety (OHS). However, there has not been a large body of literature that uses these concepts: OHS has traditionally drawn on ergonomic, psychological, and business-oriented frameworks which have not produced critical bases for evaluating the causes of occupational injury and ill-health, nor explained the broader social structure within which OHS is negotiated.

A Marxist framework centers on the belief that the capitalist mode of production is based on the production of value, surplus value, and capital, and the reproduction of conflicting relations between the classes. Marx identified issues of ownership and control of the labor process as central to an understanding of capitalist society.

Hegemony helps to explain why the benefits of control for management can still seem fair, reasonable, and perhaps the only way that industry can be organized in modern society. The dominant rhetoric in capitalist society encourages a belief in the right of management to control, and this is espoused by management and generally supported by government. Even though there is little evidence of successful needs satisfaction and fulfillment for lower classes, the belief in satisfying personal needs through meeting the needs of capitalism is dominant. Gramsci's (259) concept of hegemony helps to explain how this process operates.

The state is the governing body in the political system empowered to regulate the legal conditions under which work is carried out, and to legislate the social and other conditions within and outside work. In Australia, the federal government, the states, and to a lesser degree the local governments establish conditions under which management and labor relations are conducted and work can be performed.

These concepts in themselves do not explain the social processes involved in the production of stress and ill-health. However, they form the basic framework for understanding the maintenance of social relations that produce outcomes in favor of owners/managers.

THE THEORETICAL CONTEXT

The economic and sociopolitical framework within which the broader process of production is carried out incurs certain social as well as personal costs. The organization of work and dominance by management maintain conditions at work that are conducive to occupational stress and ill-health. However, occupational ill-health is likely to be most prevalent among blue-collar (working-class) workers (e.g., 133, 134): they are at greatest risk in that they work chiefly in order to fulfill goals established out of management's interests and not necessarily their own. These interests are bound up with management maintaining its own control.

The assumption that our state of knowledge, statistical inferences, and prevalence reporting present an accurate picture of the nature and extent of occupational stress needs to be questioned. Biases, Elling (216) argues, are class based and are basic characteristics for understanding OHS. In fact, the processes of gaining information protect owners and managers from some greater level of responsibility for the production of stress-related ill-health. Elling argues that the "day to day OHS surveillance and reporting systems . . . the problem of clinical diagnosis and reporting or lack of reporting of OHS disease . . . and the workers right to know form a knowledge base in support of ruling class hegemony" (26, p. 95).

Stress is an important area to include, as it has been part of the academic debate over the past two decades or more and has been acknowledged as a contributor to ill-health at work (7, 104, 133, 134). However, stress has been problematic in lacking a recognized and agreed cause in the workplace and has been difficult to use as a basis for compensation.

The symptoms of occupational stress have been with us for a long time, but stress is a relative newcomer as a cause of serious concern for workers' health, productivity increase, and effective management in industry. It also remains in the contentious position of being parceled away by some and being dealt with by reform that focuses on basically ergonomic solutions. Given the continued level of high unemployment, occupational stress may remain one of the costs that employees and the broader community have to bear unless effective mechanisms for identifying stress, dealing with it in the workplace, and ultimately providing compensation for it are established. The real dollar costs of stress have not been measured much beyond workdays lost, accidents, and compensation costs.

An increased interest has come about partly because of an increased occurrence of stress. Medical practitioners, government agencies, unions, and industry leaders have also become more aware of the incidence of occupational stress and the implications for employees' health as well as productivity levels. Major impacts have resulted from changes in legislation on OHS and in compensation laws.

SOCIOPOLITICAL ACTIVITY AND OCCUPATIONAL STRESS

Elling (216) presents a class-based analysis of occupational health and safety. He argues that the hegemony associated with many specific hazards leads to their meanings already having been defined. The causes of occupational stress remain rooted in a variety of management practices. Management's definitions need to be questioned. Further, he argues that OHS research tends to serve the interests of management and owners. This further strengthens ruling-class hegemonies in the consequences of work organization on stress. In Australia, for example, the emphasis on psychologically based approaches in stress research has preferred management-centered rather than employee-based solutions to problems. Research has favored individual explanations of stress at work rather than broader structural, social, and political explanations.

It has long been established that exploitation takes place in the interests of the ruling classes (see 237) and that the health of large groups of the population is subservient to those interests. Elling (216) highlights that OHS problems are more pronounced in underdeveloped countries; however, the principle of subjugating health to the interests of the major goals of profitability is evident.

Marx (237) argued the need for owners to maximize income and productivity to serve in capitalist markets. Elling (216) further argues that labor is distinguished from other commodities in workers' ability to cooperate or resist in their exploitation, and how this is influenced by the broader context within which it operates.

Burawoy (1), a sociologist, draws on the idea of a political economy and argues that a management-controled labor process is embedded in the workings of the broader political economy of a capitalist society. His analysis could be applied to occupational stress as a structural phenomenon. However, his argument has certain drawbacks and will be reviewed in the context of the more recent studies of the state.

Occupational stress should be seen as a by-product of that broader sociopolitical activity. Stress is part of a wider problem of owner and manager control of the labor process. The effects of Frederick Taylor's (3) scientific-management-controlled production processes have been a deep-seated problem for occupational stress. The issue of control of the labor process needs to be fundamentally questioned.

When management control leads to alienating work and employees fail to satisfy important needs, stress is likely to result. In addition, when sociopsychological resources (self-esteem and mastery) are low, employees may not be able to deal effectively with undue demands or with unsatisfied needs. Ill-health is a likely result. For those in the lowest occupational positions, where influence is least and there is little agreement with the goals of management, the experience of stress and ill-health is likely to be the greatest.

When management exerts substantial control at work, the ability of workers to influence or change the managerial imperatives upon which the capital-controlled labor process is based is crucial to occupational ill-health. The sociopolitical processes within which work is carried out—or as Burawoy (1) notes, the conditions of the political economy—establish the preconditions and structures for managerial control of the labor process. This exercise of control normally disadvantages lower level workers. A precondition of management control in capitalist society is a degree of worker ill-health (see 8).

Occupational stress is a consequence of the conflict of interests in industrial society: it is part of the risk imposed by the process of capital accumulation. At the broadest level, stress results from management's control of the labor process, and in particular from management establishing its prerogative to pursue its interests. It also arises from conflicts inherent in a work system that does not properly sustain the needs of workers.

Stress has only relatively recently become recognized as a legitimate basis for claim as an occupational illness or injury. In Australia, stress has been recognized as a basis for compensation claims for some time. In the past stress was largely treated as one of a number of psychosomatic illnesses. Stress, and conditions such as OOS (occupational overuse syndrome), still remain highly contentious issues in compensation courts. The conflict of interest between management, the corporate sector, and the claimant involves an interplay between medical practitioners and insurance companies. Underlying conflicts between the ideologies of capital and labor are played out in the assessment centers and compensation courts.

However, the process of getting stress cases to compensation court involves substantial inconsistencies and conflict. And bias in approaches to health care by dominant medical professionals can redirect attention from some of the basic causes of illness. Bias in health care approaches can also play down and refocus interest away from fundamental conflicts of interest between capital and labor.

OCCUPATIONAL STRESS: A DIVERGENT CONCEPT

A great deal has been written about occupational stress. Public knowledge and legislative interest have increased during the last decade in Australia and in many other western countries. However, unless the causes of stress are seen to lie in the structure of work relations and in broader class relations, occupational stress runs the danger of being treated as an individual phenomenon. Management control and the process of management that develops out of that control are a major cause of stress. This is an outgrowth of the Marxist idea of the domination of the working classes by the owners (and managers) of the means of production. That domination leads to poorer levels of work experiences, greater stress, and ultimately worse levels of health for working-class workers (see 133, 134). The relatively underdeveloped attempts to explain the phenomenon of stress and ill-health through links with the sociopolitical structure of the broader society are of interest.

Sociological writings on stress have shown the importance of social and structural factors, adding an important understanding to biophysiological and psychological models of stress. The degree of control and influence experienced by individuals is an important factor in determining the degree of stress experienced.

THE STATE AND THE LABOR PROCESS

The role of state intervention in the labor process has been largely neglected by industrial sociologists. A sociological explanation needs to account for the state's role as mediator. Burawoy (1) suggests that political mediation in the labor process takes several forms. He argues that while the state is distinct from the labor process, it establishes the parameters and guidelines for the struggle between capital and labor. There are many legislative means through which this is done. These include employment and retraining policies, establishment of protective tariffs and trade agreements, and policies on employer responsibilities for compensating for retrenchment. Mayhew and Peterson (217) have addressed many of these issues in the Australian context.

In contrast to Burawoy's sweeping analysis of the state's role, Elling (216) presents a view that shows a varied and somewhat limited role of the state. He argues that the state has a role in intervening if problems are created through capitalist exploitation to achieve greater profitability, and that state intervention has helped to avert the growth of workers' movements. He refers to state rearguard action to protect ruling-class interests as much as possible (216, p. 14). Therefore the state cannot be regarded as an impartial participant in OHS, but rather as more interested in supporting the position of management.

Major writers on the labor process give different emphases to the role played by the state in influencing the labor process. While Burawoy treats the political apparatus as distinct from and causally independent of the labor process, Littler and Salaman (162) treat the labor process as part of and consequent on the state. Braverman (2) largely ignores the role of the state. Marx (237) regarded the basis of management control as coercion in transforming labor power, while the state plays the role of presenting a set of conditions in which workers' dependence on the sale of their labor power is mediated. However, Elling (216) argues that the state's role is often to mediate on behalf of labor only when greater interests of the ruling class may be under threat.

Littler and Salaman (162) rightly suggest that mediation by the state in the labor process does not occur at only one level of work. The state establishes guidelines between management and labor interests at three levels. As Littler and Salaman suggest, this represents political mediation on the scope of managerial control (see Chapter 4). First, at the level of job design, the state establishes policies that set certain parameters on job design, especially ergonomic restrictions to ensure a minimum level of quality in terms of safety. In certain occupational groups such as craft apprenticeships, minimum skill requirements and training are required. Second, at the level of management control, decision-making, and policy creation, government authorities have established special employment and retraining schemes. Third, government legislation on acceptable lengths of working days, the working week, and holidays has long been established. Much of this state legislation has resulted directly from union activity and has required long struggles.

Apart from some legislation that ensures no undue workloads, there is little policy influence over management style. At the level of the employment economy, government policy influences the levels of employment and job security in particular industries. This is often achieved by encouraging growth in some industries and decline in others, and by regulating competition between industries through economic policy. Therefore some industries suffer high levels of job insecurity and unemployment as a consequence of government fiscal planning and policy. Australia is a long way from the ideal of a full-employment economy, which was a stated policy in the 1950s. In the 1980s and 1990s economic recovery, stability, and corporate growth have been the catchwords of industry and government. There has been some government initiative for retraining and redeployment of skills, but individual workers are now left to achieve and maintain their own security. Provision of employment as a right is not on the political agenda.

This analysis is not meant to suggest that the state impinges positively only on the position of the worker. Most capitalist governments expound the ideology of capital. With the incoming Labor government in 1983, Australia saw some changes in rhetoric about the rights of management to control. The changes, however, were only marginal. Rather than an overt identification with capital, the state under a federal Labor government expounded an ideology of necessity: that is, labor's position can only improve if it shoulders the burdens of high levels of unemployment and lower levels of wage purchasing power in the pursuit of increased productivity and profitability. The hegemony is based on the state's belief in the need for management control of the labor process. If workers can satisfy the interests of capital, this will enable them to satisfy their own needs for employment and security and at the same time will help to bring about an economic recovery. Under the recent Labor government, certain legislative changes have been less obviously supportive of management, but the question remains: does the state really protect and promote the interests of labor? The current federal Liberal government, however, is dismantling many of the mechanisms that created a cooperative working arrangement between government and labor.

The 1983 Accord on Economic Policy, a bipartite agreement between the ALP (Australian Labor Party) and ACTU (Australian Council of Trade Unions), was central to the Labor government. Capital, labor, and the state are often seen to have mutual interests, and the Labor government appealed to these interests as the basis of its economic policy.

The basis of the Accord was income and prices policy. Some unions changed their position significantly in order to meet the demands of the Accord. Stilwell (261) argues that the Amalgamated Metals Foundry and Shipwrights Union, having established a proposed people's economic program in the 1970s, acquiesced to the Labor party's call for voluntary wage restraint and supported the Accord in order to ensure future jobs for its members. Part of the ALP's success was in getting unions to agree to wage restraint; in the Accord, unions were given undertakings by the government of employment stability and growth in real wages. As a result, much of the industrial conflict prior to 1980s was minimized.

Reforms agreed to in the Accord did not completely transpire, however, and while conflict between labor and capital was minimal, labor's rewards for the agreement were not as expected. Wage restraint has not been effective because the real wage value has declined over the past few years. Wage parity was also not maintained (part of the Accord's incomes and prices policy) between managers and workers: class differences as evidenced by income and purchasing power of wages increased. This is largely because agreements were struck between management and labor elites and, while workable, still failed to meet the needs of labor such as job security and purchasing power of wages.

According to Stilwell (261), the Accord of 1983 stated the following goals. First, there should be a reestablishment of full employment as the basis of economic recovery achieved through prices control and strengthening of the Trade Practices Act. The idea of a full-employment economy was not new. The post–World War II Labor government prepared a comprehensive White Paper which enshrined the notion of a full-employment economy as a way of bridling capitalism. The Hawke government maintained that with an effective fiscal policy, the economy would be competitive in the international marketplace. Second, there should be centralized wage fixing and cost-of-living adjustments, with real wage increases. Third, for nonwage incomes there should be a strengthening of taxation and companies legislation. Finally, tax and government expenditure would shift the burden from low to high income earners. Later in 1983 a tripartite agreement between government, labor, and capital was established through a National Economic Summit. Commenting on this, Stilwell says of the incorporation of the interests of employers and financial groups that "both of these considerations . . . have been responsible for its [the Accord's] subsequent selective application, modification and re-negotiation" (261, p. 11).

The summit ignored the Trade Practices Act and modified the idea of price control. Also, no real benchmarks were established for wage fixation. As a result of the summit, election undertakings were ignored and the government did not follow through on fundamental interests representative of the labor movement. One scenario was that the left wing of the ALP was virtually disbanded. As Stilwell (261, p. 13) points out, the result of the summit was to paper over some fundamental conflicts and contradictions.

This narrative shows that management's control over the labor process is enshrined in the influence that management has over the state and its policy creation. In summary, there are three aspects of state intervention between the interests of capital and labor. While the state mediates at times between capital and labor and this affects the labor process, under the Accord management and labor maintain their conflicts of interests and at times the state can act on behalf of labor to moderate the influences of management control. However, the experience of bipartite and tripartite agreements has been that management's right to a controlling interest over the labor process is supported by government activity. A major outcome for labor has been that its bargaining base has been eroded. In a sense, the Accord diverted attention from real class differences that are reflected in labor and management differences.

Since 1996 Australia has had a conservative federal government which has revised many of the agreements upon which OHS policy had been formulated and thus has affected many of the cooperative agreements upon which the previous Labor government had based OHS reforms.

The Liberal/National coalition identified reform of Occupational Health and Safety in Australian workplaces as part of its platform. They have traditionally opposed unions and set about to weaken union activities. The unions under the previous Labor government had enjoyed a largely cooperative industrial relations environment and approach to occupational health and safety, as previously out-lined in this Chapter. The incoming government's initiatives worked toward unpicking award structures and weakening conditions and protections for workers across the country. This has been particularly evidenced by awards being weakened in many industries and the growth of unregulated and unprotected employment (217). Worksafe, now the National Occupational Health and Safety Commission has been stripped of many of its staff and resources and now under-takes only a small research function. Grants for workplace research have also diminished or been discontinued.

The Australian Council of Trade Unions mounted a national stress campaign late in 1997 and surveyed over 10,000 workers, finding that overwork, job insecurity, conflict with management, and a lack of effective occupational health and safety were the major causes of stress cited at work. This indicated a marked shift over the preceding years where overwork has become a major problem for employees at all levels, not just for management. In addition conflict with management has become evident as employees are being forced into short-term contracts and insecure work environments.

The new coalition government has pursued a waterfront reform program, per-ceived by many to involve destroying the Maritime Workers' Union. In an initia-tive reported by the union as a conspiracy between the government and a private employer, and a ruling by the Federal Court acknowledging the possibility of conspiracy, a maritime employer sacked 1400 workers who were replaced by a non-unionized workforce. This led to the court ruling the unionists' reinstatement: a series of legal initiatives were then pursued by the company, supported by the government.

The cooperative approach to industrial relations and occupational health and safety that was established through the Accord of the Labor government was quickly eroded by the coalition government, adopting a traditional platform of workplace confrontation. This has led to the eroding of many workplace condi-tions, and for conditions like stress which, due to the economic and political environment is becoming heightened, it has meant that its identification, regula-tion, and compensation has become a more difficult issue.

THE LABOR PROCESS, THE STATE, AND STRESS

We need to consider several points that link a labor process approach and stress and consequent ill-health. These involve, in the first instance, a consideration of the consequences of the capitalist labor process for the organization of work and work relations. In addition, there is the need to review the state's role in legislating conditions that set the parameters for manager/owner and worker influences on health issues at work. This includes legislating to set limits on management prerogatives, and providing compensation to workers. Finally, the competition for control of the labor process and how this has influenced the establishment and maintenance of health standards in the workplace needs evaluating. A look at the interaction of competing interests between capital, labor, and the state can show how acceptable occupational standards have developed.

Thompson (238) argues that two trends away from a Marxist labor process were greater opportunities for tighter control through Taylorist principles of work organization and developments in science and technology leading to widespread deskilling. These elements became focal points of Braverman's analysis. Lane (262) has suggested, however, that with technological change, values and attitudes from education and training can be shaped for both employers and workers. These can influence management approaches to using labor. Willis argues for Braverman that for the working class . . . the task is to find an employer who will engage their labour power potential. "For the employer the task is to transform the labour power potential into actual labour in order to provide wages and make a profit" (241, p. 7). Braverman reestablished a focus on the element of control, and Willis suggests that he reoriented the sociology of work toward traditional interest in political economy.

A number of writers in the labor process tradition have examined the effects of the organization of work and control on producing stress. These include Coburn (4), Garfield, (5), Navarro (6), and Schwalbe and Staples (7). A labor process analysis of stress and ill-health also looks at the effects of broader structural relations on work. These studies emphasize labor's lack of control over the labor process as a principal cause of stress and ill-health (see 4–7). Labor process theory argues that the structural framework within which the process of production is carried out maintains conditions at work that are conducive to occupational stress. Management's attempt to exert control over the labor process leads to alienating working conditions in which those with least influence fail to satisfy important needs or have insufficient resources to deal with excessive demands. These workers are likely to experience stress and ill-health.

It is important to consider the current climate of occupational health and how it has developed. In Australia during the last decade, moves were made toward establishing a more coherent and cohesive OHS policy than had existed both nationally and at the state government level. In Australia, the past federal Labor government established a National Occupational Health and Safety Council in

1985 which represents employers, unions, and the government. States also enacted revised legislation, including the Victoria Occupational Health and Safety Act in 1985.

Australia was well behind major western countries in introducing OHS legislation. The introduction of more recent legislation arose mainly from the labor movement's increasing interest in OHS issues and with the development, at Lidcombe in New South Wales, of the first Workers Occupational Health and Safety unit. The election of a federal Labor government and several state Labor governments saw some of these interests of the labor movement incorporated into reasonably progressive legislation in the 1980s.

Some of the greatest developments in the area of occupational health have occurred in New South Wales and Victoria. In New South Wales, joint worker-management committees were established to preside over health issues, and in Victoria health and safety representatives with some effective evaluative role are elected from trade union groups. While these developments appear progressive, Pearse and Refshauge (263) were critical of the government's slowness in legislative response to pressure from the labor movement.

Issues relating to OHS can best be understood by considering the interaction and struggle for power between capital and labor. Government has also had a vital mediating role in occupational health developments. Quinlan and Bohle maintain that the state has responsibility for a great deal of the public health arena (264, p. 81):

> as well as the infrastructure specifically addressed to OH and S [occupational health and safety]. . . . This infrastructure determines the standard and accessibility of health care for injured workers, the specific OH and S knowledge and training of health care professionals, the availability and nature of rehabilitation, and a host of other matters.

Biggins (265) has argued that OHS has long been neglected in Australia. Unions demanded action and there were a number of reasons for newer, more radical reforms during the 1980s. Few sociological writers consider these in evaluating developments in stress and OHS legislation. Pearse and Refshauge comment aptly that a realistic appraisal of health and safety activity must be seen politically (263, p. 647):

> The extent of reform in occupational health and safety in Sweden was dependent on the strength of the labour movement. In Australia this proposition has been proven to be true because the most significant reforms have occurred only when the political party of the labour movement has been in power.

They suggest, however, that the type of reform was influenced by political variations in the union movement (263, p. 647):

> in Victoria, where the labour movement [was] more to the left, greater
> emphasis has been placed on the role of the trade unions and workers health
> and safety representatives. In New South Wales, the labour movement [was]
> strong but dominated by the right, which [placed] less reliance on workers
> and trade union initiatives in the workplace.

Occupational stress has yet to achieve a high health priority. Some major legislative improvements have taken place in OHS, but occupational stress has not yet passed through the same process as more conventionally defined medical conditions and has therefore not generally achieved the same priority for compensation claims. Worksafe, the national Commission for OHS in Australia, was sharply reduced in numbers by the 1996 Liberal government and has been refocusing on national surveillance and standards. However, a more management-oriented approach toward OHS has been adopted by the more conservative Liberal government.

The role of the state has provided some points of conjecture among writers. Burawoy rightly argues that the state influences both market forces and the controlling role of capital: "first, social insurance legislation guarantees the reproduction of labour power at a certain minimal level independent of participation in production. Moreover, such insurance effectively establishes a minimum wage . . . constraining the use of payment by results" (1, pp. 125–126), and (1, p. 126):

> the state directly circumscribes the methods of managerial domination which
> exploit wage dependence. Compulsory trade union recognition, grievance
> machinery and collective bargaining protect workers from arbitrary firing,
> fining and wage reductions, and thus further enhance the autonomy of the
> reproduction of labour power.

This does not mean that the state acts independently of other influences in the community. Political pressure can be placed on the state from owner and management groups that represent certain class interests. Legislation to resolve conflicts of interest in the community as well as intraparty conflict determines the nature of the political apparatuses that protect and represent the interests of workers. Kivinens (254) argues that a new type of middle class derived from Marxism has formed, based on core and marginal groups in the middle class; marginal groups contain a number of working-class features.

Occupational stress should be viewed as an outcome of interaction between the workplace and the sociopolitical system. As Burawoy correctly states, an interaction occurs between the workplace and some broader social and political processes: an understanding of the politics of production "aims to undo the compartmentalisation of production and politics by linking the organisation of work to the state. . . . The process of production contains political and ideological elements as well as purely economic moment"; and the process of production extends beyond the labor process: "it also includes political apparatuses which reproduce those relations of the labour process through the regulation of struggles" (1, p. 122). Occupational stress is produced and sustained by the broader sociopolitical process. In brief, control of the labor process by capital represents the interests of managers and owners in capitalist economies, the managerial perspective. This perspective contends that occupational health is subordinated to the pursuit of increased productivity and profitability. Quinlan argues more insistently that occupational ill-health is one of the foundations of the capitalist system, in which "the successful subordination of worker health was an important foundation for capital accumulation and not simply an unfortunate by-product of it" (266, p. 201).

The issue is complex because it involves the interrelationship between the state and the labor process. It is a question of what Burawoy calls the politics of production—that is; it involves the sociopolitical and economic framework within which the process of production is carried out. This maintains conditions at work conducive to occupational stress. Quinlan argues that conflicting goals and interests of labor and capital place workers at risk, and states that the "basic conflict of interest within industrial . . . societies [that places] workers in particular at risk is clearly important to a broad sociological understanding of occupational illness" (266, pp. 179–180). Further, he maintains that OHS can be seen either as an occasional industrial issue or as an inseparable element of the industrial relations mosaic, and is "an integral part of the experience of people who work, a continuing part of this relationship to their job" (266, pp. 141, 144) and other experiences relating to other workers and managers. Labor's role in exercising control over the labor process has been reduced considerably over the past 70 years. Increased control by management has had a number of consequences First, for management, a major problem has become how best to organize labor. Second, as Braverman and other labor process writers show, labor has experienced a marked decrease in its control over the production process. Third, workers have experienced reduced skill levels in certain jobs; and finally, there has been an increasing emphasis on creating and maintaining class differences between jobs. In a work system where management exercises the right to dictate the production process, many negative work experiences are accepted. Unions are increasingly opposing management's right to create and maintain health risks for workers in order to achieve its goals, but it is characteristic of a capitalist labor process that workers will accept the responsibility for their own stress and ill-health. The acceptance of managerial domination, however, certainly poses problems for the notion of a class consciousness among workers.

Willis (260) argues for the need to consider a labor process approach to issues of OHS. He maintains that the pursuit of productivity and profitability is a major management goal in the process of health and safety at work, when (260, p. 321):

> legitimated by managerial prerogative, concerns about occupational health
> and safety could, and have, traditionally been subsumed under production
> goals.

He argues that the tendency has been for ill-health at work to be individualized by the dominant class and backed by the state and that this leads to "mystifying its structural causes" (260, p. 321). Willis does suggest, however, that there has been some important softening of legislation and of the attitudes and role of trade unions which, among other reasons, has led to a growing emphasis on prevention of ill-health rather than compensation.

The labor movement has become more organized in order to combat the causes and effects of occupational stress, but it has been slow to take on the issues. While a lot is known about the causes of stress at work, the union movement has only quite recently been interested in stress as an issue. The Trade Union Handbook for Health and Safety Officers (267) states that it is a union responsibility to work toward a stress-free work environment. Teachers' unions in Victoria are more aware of stress in their work, particularly stress caused by structural issues in schools and the education department. This increased awareness is largely due to Rosemarie Otto's (18) studies of teachers. Largely as a result of her work, teachers have become one of the most active occupational groups in fighting stress compensation claims.

HEALTH AND THE CAPITALIST LABOR PROCESS

According to Doyal, capitalism's foundation is the subordination of the health of workers to its own goals (236, p. 44):

> under Capitalism there is often a contradiction between the pursuit of health
> and the pursuit of profit. Most attempts to control the production of ill-health
> would involve an unacceptable degree of interference with the processes of
> capital accumulation, and as a result, the emphasis in advanced capitalist
> societies has been on an after-the-event curative medical intervention, rather
> than broadly-based preventative measures to conserve health.

Schatzkin (8) argues further that the concepts and ideologies of both health and medicine are contained within the social relations of production. These ideologies incorporate the notion of the dominance of medical expertise. The dominant medical ideology supports and reinforces the belief that health is an individual responsibility. This belief is maintained to further capitalist interests. Dominance by the medical profession and its support of the capitalist system is well documented by Willis (268).

Schatzkin (8) rightly notes that health in capitalist society is seen as the ability to work for those who own the means of production. Capitalists are not concerned with workers' health beyond what makes them capable of maximizing productivity. Workers, of course, are interested in their own health in terms of increasing their quality of life. As Schatzkin observes, health as labor power is "the tendency towards maximisation of the rate of exploitation" (8, p. 217). Not only do workers suffer as a result of the capitalist production process, but management's control over the labor process necessitates this in the process of accumulating surplus value. The capitalist class and its level of health are dependent upon the exploitation of the working class. Schatzkin argues further, "so long as the Capitalists hold power and are the dominant class, health as labour-power rather than as workers' quality of life will best correspond to the underlying structure and function of the political-economic system" (8, p. 217).

The greater the value of workers for increasing productivity (i.e., highly skilled and specialized workers), the greater is the need for owners/managers to ensure their health and survival. Workers who are easy to replace (i.e., unskilled workers) are not regarded as investment material in relation to maintaining their health and therefore productive capacity. Schatzkin (8) aptly argues that the main aim of health care is to exploit labor: in an economy where unemployment is high and workers are easily replaced, health and medical care decline. The health of those who have least power and influence in the productive process is most likely to be affected by the capitalist labor process.

Occupational stress and ill-health have long been treated as the responsibility of individuals and as the result of individual action. Increasing attention has been given in sociological writing to the social basis of ill-health: that is, those in lower classes in capitalist society are most susceptible to ill-health. Mathers (235) and others provide evidence that many causes of death are more prevalent amongst lower socioeconomic groups.

Medical practitioners form a professional class that maintains and substantiates its position of control over those controlling the production process by acting to promote the hegemony of the ruling classes. Illich (269) argues that the medical profession acts outside class boundaries, manipulating medical knowledge and practices to support and enhance its own power. I support this notion and argue further, however, that the medical profession's role and power need to be seen in the context of maintaining and enhancing class power differentials.

Willis (268) aptly argues that the medical profession forms a relatively new class group of professionals who tend to support and legitimate the interests of capital: they represent an ideology of dominance. As noted earlier, underlying conflicts between the ideologies of capital and labor are played out in the assessment centers and compensation courts. Quinlan (266) observes that bias in medical care by health professionals can refocus the basis of illness or injury. Further, the dominant notions of health care obscure a fundamental conflict of interest. Quinlan argues that "the suffering of workers is reinterpreted and even distorted by medical diagnosis and treatment" (266, p. 198). The role of the medical profession in occupational stress cannot be adequately understood outside the boundaries of the compensation system. For a number of conditions, medical research, epidemiological studies, and other data surveillance are not regularly conducted in workplaces. Consequently there is a great deal of occupational illness for which etiology and diagnosis are not well understood.

The medical profession has characteristically treated stress as a psychosomatic complaint, and in compensation cases it most often represents the interests of management. It often plays a watchdog or punitive role with workers. Victims of occupational stress and ill-health are often treated with suspicion in compensation medical examinations. Comcare provides very useful information in defining stress and offering ways of dealing with it. Generally, however, considerable effort is needed by state bodies to inform the community about stress in the workplace, to provide claimants with information and guidance about stress, and to provide employers with useful guidelines and information.

The medical model can lead to medical ill-health being treated as an individual complaint and often as the fault of the individual worker. The medical model needs to change substantially in order to provide effective diagnosis and treatment for a large number of conditions.

Dawson (270) argues that the compensation courts are supportive of management's suspicious attitudes toward individual problem workers. In compensation court, the onus is on the individual worker to prove the cause of illness beyond any reasonable doubt. Both the legal and medical systems place individual claimants at some disadvantage. This supports Figlio's (271) claim that the medical profession is also intimately tied in with the social insurance interests in its definition of occupation ill-health symptoms. In this sense the medical profession mirrors the dominance of control by capital and perpetuates the hegemony that individual workers serve their health and other needs by subordinating their own needs to the interests of capital.

THE ROLE OF THE STATE IN RELATION TO STRESS

An important part of the sociological explanation of stress is preventive and compensatory legislation by the state. This establishes guidelines and sets some limitations on managerial responsibility in capitalist society. The system can also give workers the right of redress. Past legislation has been aptly described by Carson and Henenberg (272, pp. 4–5) as supporting an ideology of dominance, and such crucial ideological effects are not just accomplished once and for all by formal enunciation through law—symbolically powerful as such enunciations and their attendant corpus of legislative knowledge may be. Rather, or in addition, it must be recognized that the constant reproduction of such important ideological representations and distinctions as natural, timeless, and to be acquiesced in is a matter of continuous ideological work carried out through the actual practice of the institutions comprising state and civil society. In the workplace, the activities are carried out by those whose interests would be best enhanced by opposition to the hegemony of the ruling class.

It needs to be seen that the state has a major role in the occurrence and consequences of occupational stress, and that the nature of its role is inextricably bound to considerations of the political economy. Compensation is important, but it does not focus on structural causes of occupational stress. Inherent conflicts over control of the labor process are at the root of occupational stress.

While legislation is aimed at preventing ill-health (including stress) and accidents, its formulation in fact obscures the basic causes. Deves (273) correctly argues that the ideology of OHS legislation requires a certain dependency through compensation, with workers thereby assuming a patient role, rather than fostering a move toward increased autonomy, control, and therefore determination over both the workplace and the labor process (see 274). Federally, the structure of Worksafe in Australia depended upon psychological interpretations of stress for compensation. This makes structural interpretations of the causes of stress, and thus structural change, unlikely under current provisions of stress-at-work legislation.

CONCLUSION

Burawoy and other writers on the role of the state have clearly shown the extent to which the state legislature and political processes mediate the effects of management and labor relations in creating conditions of stress and ill-health. However, Elling (216) shows through an analysis of OHS legislation across a number of countries that the role of the state varies and can be shown to form a set of typologies of intervention. This broader consideration of political and social processes helps to explain the operation of a system of work, which produces stress and ill-health differentially among occupational groups, regulates its occurrence, and yet provides ideological justification.

Occupational stress and ill-health occur as part of the context of the capitalist system of work relations. Quinlan (76) argues that occupational ill-health is a necessary outcome of the capitalist mode of production. A sociological understanding of stress requires, first, that it be understood as an outcome of broader sociopolitical relations. In Australia, government legislation regulates class relations both inside and outside the workplace. Second, sociocultural factors help to explain why stress and ill-health are experienced more by those with little opportunity for influence and control. Labor process writers show that an understanding of the sociocultural and political context clarifies the role and process of management control of the labor process as a cause of stress and ill-health for workers.

The state plays a number of roles. Writers such as Burawoy show that it establishes and preserves class differences through government policy that maintains ideological differences between the classes: in regulating the capitalist labor process, the state strengthens differences between owners, managers, and workers. However, it also mediates on behalf of labor in creating and administering certain favorable conditions at work, and in OHS legislation. The Australian Labor government, through the 1983 Accord, established a regulation of class conflicts and of conditions over which control of the labor process could be enacted. In particular, federal and state Labor governments' OHS legislation has had a marked effect on recognition of stress in the workplace and on legitimating its compensation. Occupational health and safety has to be seen as part of a struggle between capital and labor. However, as Pearse and Refshauge (263) argue, it realistically needs to be seen politically also. Some sociological writers have aptly observed that the recognition, treatment, and compensation of occupational stress are first and foremost a political and ideological issue.

The reasons for linking labor process, the state, and occupational illness are many and complex. Quinlan (264) argues that simply viewing occupational illness in the workplace is too narrow a focus and does not look at the range of factors involved in the explanations of occupational ill-health. He argues that (204) explanations of occupational illness in the workplace or industry fail to address the broader social nature of illness and health. Studies fail to explain why more demands for safe working conditions have not been made. The dominance of medical and health explanations provides a strong explanation for this.

Quinlan further observes that intervention by the state has had important effects. And class relations played out through worker and management struggles have also ensured a less than safe working environment. The concept of a political economy enables a broader view of stress as an outcome of the fundamental social and political workings of the capitalist system.

CHAPTER 6

Design of a Study on
Stress at Work

The purpose of this chapter is to clarify the major concepts used in this study of stress and ill-health, to discuss issues relating to the research methodology employed, and to outline the major dependent variables used and their conceptualization and use in the study. I discuss the interrelationships between the major variables and present issues relating to the use of a survey questionnaire and issues that arose during data collection.

A number of research approaches were possible to study stress and ill-health at work. A qualitative study based on open-ended interviews would have provided insights into conditions at work leading to stress, and to some of the pressures likely to be conducive to ill-health. However, a quantitative study based on the use of an extensive questionnaire allowed causal connections between important variables in the stress process to be more easily identified and measured. A very large number of studies of occupational stress have used quantitative methods employing established measurement tools. While this does not preclude the use of a qualitative research design, the use of a questionnaire employing a number of validated measures was preferable. For example, it enabled direct comparisons between this study and a number of other studies of stress and ill-health. And a greater number of subjects could be included in this type of study than in an interview-based program, particularly as a large number of variables were being measured. Finally, the questionnaire allowed the use of a number of validated measuring instruments (scales) that could be compared with their use in other studies.

A number of hypotheses were tested during the study, and further tests were based on expected results derived from mainly quantitative studies reviewed in Chapter 2.

Certain information collected on job satisfaction, other social support, and some other contextual factors was not included in the final analysis.

I selected a research setting that would allow for an intensive investigation of causal factors in the occupational stress and ill-health process. The investigation involved measuring the nature of work experience for workers by questionnaire, and measuring a number of dependent variable outcomes (including stress, low self-esteem and low mastery, and somatic and psychological ill-health). These measures have been developed with a model of relationships in the stress and ill-health process based on an evaluation of the literature reviewed in Chapters 1 to 4. The study of stress treated each element in the process as causally related to the broader sociocultural and political context of an advanced capitalist society. The experience of alienating work that results from limited control over the labor process forms a vital link in this conceptual framework and design.

The research is based on labor process theory of the determinants of stress and ill-health. Labor process writing has shown that blue-collar workers are most likely to have the most negative work experience, including the least control over work and the greatest lack of opportunity for using skills on the job. They are also most likely to experience the greatest stress and ill-health. Writers such as Braverman (2), Navarro (6), Garfield (5), Coburn (4), and Schwalbe and Staples (7) argue that control is exerted over lower level workers: labor process research has shown that blue-collar workers have more highly alienating work, higher stress, and poorer levels of health than other occupational groups as a result of experiencing less control. A small group of writers (see Chapter 2) have also shown that women are likely to experience some disadvantage in negative work, stress, and ill-health compared with men.

The labor process debate identifies sources of negative work experience, especially a lack of control over the job and a lack of opportunities to use skills. These variables have been developed for the analysis: they are outcomes of a division of labor and deskilling (see 2, 238, 244). Importantly, writers on occupational stress have also identified other aspects of negative work (as reported in Chapter 2). A lack of appreciation and consideration by management, role conflict, poor relationships, and overwork are additional factors developed from these writings. These factors have been developed in order to test other labor process writings by Littler and Salaman (162, 242) who argue that these factors form aspects of control over structure and the climate of attitudes and values in the organization. Based on their work, two factors, a lack of control and a lack of skill discretion, measure aspects of the task itself, while the remaining four negative work factors measure broader aspects of labor process control—that is, control over structure and climate.

Further labor process studies such as those by Coburn (4), Garfield (5), and Schwalbe and Staples (7), as well as a large number of studies reported in Chapter 2, show that when workers have insufficient control, stress is highest. Many studies have shown that a high level of stress can lead to ill-health. In addition, workers in blue-collar occupations are most likely to experience a lack of control and, as a result, are likely to have the poorest levels of self-esteem (7): this in turn is known to lead to ill-health. Other studies (see 178) have shown that a sense of low personal control and mastery can also contribute to ill-health. These three sociopsychological variables have been tested in previous research (see 178) as together contributing to ill-health.

As labor process writers have shown, the effects of control can be derived through occupational membership. A small number of studies have also indicated that these effects have a lesser effect through gender (see Chapter 2). Both of these groups are tested to determine their relative effects on all dependent variables.

The stress and ill-health models used in this study look at work demands and the inability of resources at work to meet important needs. Also examined are the effects of these demands and lack of resources on low self-esteem and low personal mastery. Stress (a negative emotional state) and a lack of personal resources (low esteem and low mastery) are then tested for their effects on ill-health symptom frequency and psychiatric impairment (as measured by the General Health Questionnaire, GHQ). The model of stress and ill-health being tested looks at the effects of alienating work in determining high stress and a diminished self-concept (low esteem and low personal mastery).

The model consists of a wide range of demographic, socioeconomic, psychological, and sociological variables. These are arranged into five causally related blocks (Figure 1).

The variables in the first block are gender, age, level of education, partnership status, and number of children. Previous research outlined in Chapter 2 suggests that age (182) may contribute to stress, while Hinkle and associates (184) and Wan (185) found a relationship between educational level and coronary heart disease morbidity. Few studies have investigated the effects of partnership or

ASCRIBED AND ACHIEVED STATUSES	OCCUPA- TIONAL STATUS	INTER- VENING VARIABLES	SOCIO- PSYCHO- LOGICAL OUTCOMES	ILL-HEALTH OUTCOMES
Gender	Occupational level	Negative work factors	Stress	Ill-health symptom awareness
Age	Years in the one position		Low self-esteem	
Educational level				General well-being
Partnership status			Low personal mastery	(GHQ)
Children				

Figure 1. The relationship between negative work experience, stress, and ill-health.

number of children, although some studies have emphasized the need to account for effects of extraorganizational factors on work stress. Gender has been found to have some effect on stress (146).

Variables in the second block are occupational position and length of time in the one position. Previous labor process research (e.g., 4, 5, 7) has shown that differences in occupational position reflect different levels of control experienced by employees, and have important effects on stress and ill-health outcomes. Caplan and associates (83) reported that the length of time spent in the one position had important effects on stress. These variables have been included in order to test the effects of control and levels of influence in the organization.

The third block consists of negative work factors. A large amount of research has been reported (see Chapter 2) on the effects of negative work factors on stress and ill-health. A body of labor process literature as well as studies by Karasek (65, 66) and others have shown that lack of control has important effects. However, a variety of other work factors which also stem from a lack of control have been shown to have an important influence on stress and ill-health outcomes. These have been included in order to test the findings of a large body of research.

The variables entered in the fourth block are stress, low self-esteem, and low personal mastery. A large amount of research, based primarily on the work of Selye (9, 15, 21), has shown stress to be an important determinant of ill-health. Research reported in Chapters 2 to 4 shows that stress is an outcome of preceding variables in the model (in particular negative work experience) and can lead to ill-health (e.g., 10, 65, 66). In addition, low self-esteem and low mastery have been found by writers such as Israel and coworkers (101) and Pearlin and Schooler (99) to have effects on ill-health outcomes. Pearlin and coworkers (178) included these variables together when examining determinants of ill-health and found that low self-esteem was a stronger determinant that poor mastery, although both had important effects.

The variables in the final block are ill-health and a psychological measure of ill-health (GHQ). A large number of studies reported in Chapters 2 to 4, including labor process studies (4, 5, 7) and others (10, 65, 66, 83), have shown that ill-health is a consequence of a process that includes preceding variables in the model. The GHQ has also been used more recently in studies of occupational health (e.g., 275–277), but few studies as yet have examined its relationship to stress. The GHQ has been used in many studies as a measure of psychological health (see 278, 279). A number of studies reported in Chapter 11 show preceding variables in the model as determinants.

For each of the dependent variables, from negative work, stress, low self-esteem, and low personal mastery to ill-health and psychiatric measures, pooled analyses are undertaken to test the specific effects of independent variables (primarily through multiple regression analysis). Specific analyses used in labor process writings are also carried out on the effects of several independent variables, as well as tests of association between independent and dependent variables. For each dependent variable, separate analyses of occupational position and gender are also carried out to test the effects of control through these variables.

The primary aim of the research is to investigate the structural relationships that contribute to occupational stress.

HYPOTHESES AND TESTS

The hypotheses tested in this study are:

1. As occupational level decreases, stress is expected to increase (Chapter 7).
2. As occupational level decreases, low self-esteem is expected to increase (Chapter 8).
3. As occupational level decreases, the frequency of ill-health symptoms is likely to increase (Chapter 9).
4. Occupational stress is expected to be an important determinant of ill-health symptom frequency (Chapter 9).

Each hypothesis is tested through regression analysis.

Apart from these four specific hypotheses, tests are carried out based on results that might be expected from literature and research: these expected results are not tested in the analysis of results as specific hypotheses subject to verification.

An overall test is on the effects of management control over the labor process on levels of alienating work for blue-collar groups. These workers can be expected to have the highest stress, lowest self-esteem, and lowest sense of personal mastery. In addition, tests are carried out on the awareness of physical and emotional symptoms and on psychiatric impairment. I also test the relationships between the degree of control experienced and ill-health outcomes.

Chapter 7 includes a number of tests, based on general hypotheses or expected results. The relationship between alienation and occupational position is tested. Previous research has shown that alienation will increase as occupation moves toward the lowest levels. Also tested are different effects of alienation of tasks, alienation resulting from control over structure by management, and effects of the climate of negative attitudes by management; the effects of occupying positions between management and blue-collar levels on alienation experienced; and differences in occupation and gender for all alienating work factors. Previous research has shown lower occupational groups to be most disadvantaged.

In Chapter 8 I test a number of effects of independent variables on occupational stress. Previous research (reviewed earlier) has shown that blue-collar workers are expected to perceive the highest level of stress, and managers and other white-collar workers to experience the least occupational stress, principally due to experiencing the least alienating work. In Chapter 9 analyses test the effects of alienating work experience on low self-esteem and low personal mastery. The research reviewed shows that managers can be expected to report the highest degrees of self-esteem and self-mastery, and factory workers the lowest levels. Supervisors, white-collar (office) workers, and skilled workers are expected to record moderate levels of self-esteem and mastery. Little research, however, has tested the relationship between levels of personal control over work and perceived degrees of personal control or mastery.

Chapters 10 and 11 examine the effects of all other variables: ill-health symptoms in Chapter 10 and psychiatric impairment levels (GHQ) in Chapter 11. Tests are undertaken in Chapter 11 to determine whether managers are most likely to report the lowest level of psychiatric impairment, and blue-collar workers to report the highest GHQ scores. Supervisors, white-collar (office) workers, and skilled workers are expected to report levels of impairment between those of factory workers and managers. I also carry out tests to determine whether workers with high stress, low self-esteem, and low mastery display the highest GHQ levels.

The study looks at the relationship between occupational position, gender, and the perception of stress and other sociopsychological outcomes in the occurrence of ill-health. This relationship cannot be understood adequately without some wider, more general notion such as alienation. As argued earlier, alienation, a Marxist concept, can be regarded as an objective, socially structured situation that arises out of management's control of the labor process (see 4, 5).

Chapters 7 to 11 test relationships in the causal model shown in Figure 1, which identifies five distinct stages in the causal path. The model principally identifies the chief determinants of stress, other sociopsychological states, and ill-health (both somatic and emotional, and low general well-being or GHQ level). Each stage in the model identifies status, events, or outcomes that are causally connected: the first events in the model are likely to have an effect on the next and later variables in the model. The first stage, ascribed and achieved status, has usually preceded any of the subsequent stages.

In all regression analyses in the following chapters, these are the first effects tested for on dependent variables. They will directly affect occupational status variables and have indirect effects through these and subsequent variables in their total effects on the dependent variables being measured. The same is true for the subsequent groups of intervening variables as they are measured for their effects on dependent variables.

In Chapter 7 the analysis focuses on the effects of independent variables (on the left-hand side of the model in Figure 1) on negative work factors. Several analyses test the effects first of ascribed and achieved status on the experience of negative work. Some of the status effects will result from indirect effects due to occupational positions held by workers and length of time in the one job. The effects of occupational membership and years in the one position are also tested for their effects on negative work. In Chapter 8 I test the sequential effects of status, occupational level, and negative work for their effects on the experience of occupational stress. Several tests are carried out to determine the strength of effects of the different independent variables. The causal effects of all independent variables are tested on low self-esteem and low personal mastery in Chapter 9. Again, several tests are made of the effects of status, occupational level, and negative work factors. In Chapter 4 I argued that stress and other sociopsychological factors should be treated as mutual outcomes at this point in the causal model. The last factors in the causal model, ill-health symptoms and psychological ill-health (GHQ) are the final dependent variables. In Chapters 10 and 11 I carry out a number of tests based on labor process research into the effect of class and occupational status on ill-health, and test labor process arguments on the role of alienating work experience, in creating stress, reducing self-esteem and a sense of personal control (mastery), and therefore affecting ill-health. The effects of all variables on the left side of the model will be tested on ill-health measures and GHQ.

COLLECTION OF DATA

I met with the director of a Melbourne stress consulting agency to discuss the incidence of occupational stress in Melbourne organizations. During the discussion she mentioned that employees from a particularly large manufacturing company had made numerous claims for stress. Many of the company's managers and other employees had also been counseled for stress by the agency. She indicated that the company appeared to be aware of many stress-related problems.

The director of personnel of the company was contacted and we began discussions about conducting a study in the company. He was enthusiastic and suggested that the whole company might be used in the study. We considered a comparative study with another manufacturer in the same region, but the opportunity to study in some depth the experience of stress and ill-health in one organization was opportune. The director offered complete cooperation and some funding to cover the cost of questionnaire printing and mailing. He agreed to feed back results of the study to specific workgroups and that these results would be used as the basis for discussion and some decision-making. The company insisted on remaining anonymous, and we gave an undertaking not to publish or submit research material containing its name. The training manager was assigned as contact person and she provided background information on the company. While she was prepared to give some descriptive information about the company, she did not make available company records on health and safety and stress claims.

The company allowed restricted access for discussions with employees. At tea breaks, lunches, and dinnertimes discussions with employees on issues relating to data collection took place freely. The company, however, reserved the right to distribute questionnaires to employees through its own channels; for production scheduling reasons, it refused to give employees time off to complete questionnaires.

We made arrangements to distribute questionnaires to all employees in the company. The company arranged to distribute questionnaires along with employees' pay over a four-week period. Wage earners received their questionnaires in the first week; monthly salaried employees received theirs four weeks later. Close to 900 questionnaires were left with the company for distribution. Questionnaires were distributed to all employee groups including sales staff at various suburban and country locations. Questionnaires were enclosed in folders with accompanying letters of explanation. Sample participants were asked to return their questionnaires in two sections two weeks apart. The first section contained two parts of the questionnaire, asking for information on work factors and their stressfulness and for responses on various health, psychological, and personal resources scales. The second section contained the third part of the questionnaire, asking for responses on a number of different coping dimensions: this information has not been used in the study. Of all the questionnaires returned, 242 were suitable for inclusion in the study, a response rate of 30%. Questionnaire recipients were not given a choice in receiving questionnaires.

In gaining cooperation from employees and management to participate in the study, completion of questionnaires was an industrial relations issue. The overall effect of a ban placed by the Amalgamated Metal Workers Union on completing "attitude" questionnaires was that a large number of members felt obliged not to participate due to union direction. We explained to some groups of employees that the stress survey was not strictly an attitudinal survey; however, union directives markedly affected responses. Also, some initial instructions by the local union representative dissuaded employees from completing the questionnaire. He had argued that it was a management survey. After we explained the study to him, he started encouraging workers to complete returns, but by that stage it was too late.

Because the study was not the result of workers' initiatives, this affected their willingness to give full support to the project. In addition, a common problem with stress and ill-health among managers is that to indicate ill-health is to suggest a weakness; that is, a manager could be afraid of appearing unable to handle the job well. In fact, the training officer for the firm said that managers in the past had actually not offered any information about stress or ill-health because they were afraid of not appearing capable of meeting the job demands.

As a result managers did not fully support the project. And at a time of large-scale retrenchments in the industry, many employees were not overly happy about participating. In a manufacturing environment where stress claims and compensation are an industrial relations issue, information that could possibly identify workers' attitudes toward the effects of management style and policy and could imply managerial responsibility might have serious implications for internal worker-management relations. Inability to gain full support from workers and management because of possibly identifying "fault" for worker health and safety is a fairly common problem in research in an industrial relations environment. Everything is seen as an industrial relations issue and is therefore defined that way.

During the period between questionnaire distribution and return we held several meetings with employees from the company at which we answered questions about the questionnaire and the study. A series of group interviews about the causes of stress at work were conducted with approximately 50 groups of about ten employees each. The interviews were informal and conducted in the company cafeteria on eight occasions over a two-week period. During the first week, we spent more than two hours on each of two lunchtimes conducting group interviews. Five evenings and three lunchtimes were spent in the cafeteria over the two-week period.

One group of older men interviewed who had not completed questionnaires insisted that management might be aided by their participation in the study. Under no circumstances, they said, were they prepared to do anything that might benefit management.

The initial return was less than 200 questionnaires. Another lunch and evening mealtime were spent in the cafeteria encouraging participants to return their questionnaires. In all, 242 completed questionnaires were returned through the mail. Some employees were very cooperative in encouraging their workmates to complete questionnaires. The editor of the company's in-house magazine at one stage published an article about the study and called for employees to complete and return their questionnaires quickly.

We took a factory tour during an evening shift to see some of the conditions under which shiftwork was done. Typical workgroup formation and organization was pointed out, together with potential hazards relating to the machine-operating work.

STUDY DESIGN

The basis for selecting a large manufacturing organization was to look closely at occupational differences in one organizational environment in order to study in depth the impact of the labor process on stress and occupational ill-health at different occupational levels. This allowed a number of variables to be controlled for, including corporate policy, the effects of industry and labor market trends, and the effects of labor process strategies. I make no claims that the study sample is a representative sample, but it does represent characteristics fairly specific to the manufacturing industry. Burawoy advocates the development of more intensive studies of single firms in order to identify more detailed implications of the effects of labor process organization. This is not a historical study, but it does identify a sample of workers from different occupational groups exposed to these labor process effects. On the value of selecting a single organizational environment, Burawoy says (239, p. 190):

> The few attempts at concrete analysis of changes in the labour process have usually emerged from comparisons among different firms. Such causal analysis, based on cross-sectional data, is notoriously unsatisfactory under the best of conditions, but when samples are small and firms diverse, the conclusions drawn are at best suggestive. As far as I know, there have been no attempts to undertake a detailed study of the labour process of a single firm over an extended period of time.

While the study was not conducted over successive time intervals, it does examine in depth the issues linking the labor process to stress and ill-health and does identify some of the historical changes in the utilization of skill levels, technological changes, and employment trends. In short, it looks at the impact of labor process organization by management on the organization of work and, in depth, at its effects on the stress and ill-health of workers.

DEPENDENT VARIABLES USED IN THE STUDY

The composition of each dependent variable analyzed in the study is outlined together with its method of measurement and the construction of scales. All composite dependent variable scales used in the study have been scored on a scale of 0–100, with the exception of Likert Scores (0–36) for Goldberg's GHQ scores. (This is discussed later in the chapter.) The 0–100 scores allow for an easier comparison of occupational and gender differences for the main scales, and an easier interpretation of unstandardized regression coefficients (b scores), which are measured as number of points out of a possible 100, therefore giving a clearer indication of the strength of the unstandardized coefficient. Scale scores have been converted by dividing total scores by the number of items in the scale, and dividing this score by the highest point on the item scale. The resultant score is multiplied by 100.

Negative Work Experience

A number of scales measuring negative work experience have been used in previous studies. The scale used by Caplan and associates (83) is extensive but does not include a sufficient range of items appropriate to this study. The negative work scale used by Schwalbe and Staples (7) is far too brief for an in-depth study investigating the influence of a whole range of effects of management control. Karasek's (66) negative work scale includes only highly specific stressors measuring a lack of skill discretion and work overload. The measure developed for this study is a composite scale developed from the above studies and a number of others (including 99).

The work factor scale contains 34 questions. It asks respondents to indicate the degree to which they perceive each of the work factors to be present in their jobs. Two questions have not been included in the negative work and stress scale in this study (the stress scale is reported later in the chapter) because of a large number of missing responses. "Do you have suitable problems finding childcare while you work?" is addressed only to respondents with children, and many employees did not respond. The question "if you have supervisory or managerial responsibility do you encounter problems in your relationships with the people who are directly responsible to you?" was also unanswered by many respondents.

Employees were asked to indicate one of the following four responses on a scale of 0–3: 0, no or never; 1, a little or sometimes; 2, a fair bit or moderately often; 3, very often or always. All items included in the negative work scale are presented with sample mean and standard deviation (S.D.) scores in Table 1.

The scale was factor analyzed and forms six separate component factors. The factor analysis method clusters groups of items each under one measure. Instead of a large number of items, there is a small number of distinct sales each measuring a discrete different property of negative work experience. The reason for grouping items into factors is that several questions measured similar aspects of the same work factor and can be considered as closely related. Alwin points out that factor analysis has been used extensively by sociologists: "It has primarily been used in the area of index construction . . . [where] the researcher is interested in determining the amount of linear dependence among a set of items or variables which presumably measure the same general domain or content (280, p. 191). Once items are grouped together they form "factors" or composite measures. A factor analysis was carried out of all items in the work scale. Mixtures of the items used in the scale have been used in a variety of studies, including those by Caplan and associates (83), Karasek (65, 66), Pearlin and associates (178), and Schwalbe and Staples (7).

We devised scales rather than using previous scales for several reasons. Other scales do not include the range and composition of work experience items suitable to a more in-depth study in a manufacturing environment. Even though several validated studies exist, a greater number and variety of measurement devices can only enhance our understanding of the causes and consequences of occupational stress. Some single items also have been included because they measure important aspects of negative work experience but have no other items clustering with them. Each factor in the factor analysis represents an aspect of job or work organization that results from management policy and reflects the managerial prerogative to influence and determine the nature of the work experience for workers: at the level of the job, the organization of work, and the climate of attitudes and values. These factors represent several types of control. Table 2 shows the results of the factor analysis.

Table 1

Work factor scale: negative work experience

		Mean	(S.D.)
*1.	Do you have variety in your work?	1.02	(.95)
2.	How often do you have a lot of work to do in a limited time?	2.03	(.81)
*3.	When there is a problem on the job, can you get help or advice?	.78	(.84)
*4.	Do you have the freedom to organize and do the work the way you want to?	.92	(.99)
*5.	How often are you told you are doing a good job?	1.86	(.79)
6.	Are you given conflicting orders at work or told to do things in ways you don't agree with?	1.03	(.80)
7.	Are there conflicts or tensions among the people you work with?	1.31	(.83)
*8.	Do you meet people at work whose company you enjoy?	.91	(.72)
9.	How often do you have a lot of noise, dirt, or dust on the job?	1.65	(1.12)
*10.	Are there times when you have little to do?	2.30	(.69)
11.	How often do you need to learn new ways of doing things to keep up with the changes?	1.42	(.81)
12.	Have you any chances for promotion or advancement?	1.84	(.90)
13.	Is there any uncertainty about your job security?	.82	(.86)
*14.	Does your work make use of your training, skills, and capacities?	1.08	(.92)
*15.	Does your immediate supervisor or boss treat you with consideration and respect?	.75	(.82)
*16.	Do you feel that you have been treated with consideration and respect by management?	1.23	(.98)
17.	Are you ever expected to do a job without adequate equipment, materials, or other resources?	1.00	(.86)
18.	Are you ever expected to make decisions without all the relevant information?	.88	(.71)
19.	Are you ever unsure of what your duties are meant to be?	.68	(.78)
*20.	Can you discuss work problems with your colleagues or workmates during working hours?	.78	(.86)
21.	How often do you work more than eight (8) hours a day?	1.83	(1.01)
22.	Do you ever work nightshift or late afternoon shift?	.50	(1.00)
*23.	Can you influence organizational decisions or policy decisions that affect you?	2.08	(.88)
24.	Have you any communication problems due to language difficulties?	.19	(.46)
25.	How often are you supposed to do more than one task at a time?	1.90	(.96)
*26.	Is your work appreciated by your immediate supervisor (in the section in which you work)?	1.14	(.91)
*27.	Is your work appreciated by management?	1.47	(.95)
28.	Do you ever take work home?	.81	(.95)
29.	On your job, how often do people act toward you as if you are a person without real feelings?	.84	(.80)
*30.	Do the people you work under ever take notice of your ideas?	1.50	(.87)
*31.	Once you leave your workplace do you find time to relax?	1.15	(.84)
32.	Does your work make it difficult to get through your chores at home?	1.12	(.92)

*Item reverse coded before inclusion in the scale.

Table 2

Factor loadings, reliabilities, and average inter-item correlations:
negative work

		Factor loadings[a]				Scale Reliability	Average inter-item correlation
		I	II	III	IV		
1.	No consideration by management						
	i. Work appreciated by the supervisor[b]	.79	−.07	.14	.16		
	ii. Treated with respect by management[b]	.77	−.05	.03	.26		
	iii. Treated with respect by the boss[b]	.74	.14	.09	−.02		
	iv. Told you are doing a good job[b]	.70	−.20	.06	.15		
	v. Work appreciated by management[b]	.69	−.20	.14	.24		
	vi. Ideas not listened to	.62	−.13	.50	.05	.86	.51
2.	Workload						
	i. Too much work in a limited time	−.02	.78	−.10	−.23		
	ii. More than one task at a time	−.11	.68	−.30	.15		
	iii. Taking work home	−.28	.61	−.16	.11	.64	.38
3.	Skill discretion						
	i. Learning new tasks[b]	−.01	−.17	.74	−.02		
	ii. Variety in work[b]	.43	−.19	.64	.00		
	iii. Work makes use of skills[b]	.02	−.21	.51	.28	.56	.30
4.	Poor relationships						
	i. Not treated with real feelings	.11	.09	.16	.75		
	ii. Tensions at work	.26	−.06	−.04	.71	.49	.35
5.	Role conflict	n.a.	n.a.	n.a.	n.a.	n.a.	n.a.
6.	Lack of freedom[b]	n.a.	n.a.	n.a.	n.a.	n.a.	n.a.

[a]Factor loadings were obtained using Varimax rotation with unities in the main diagonal. Eigenvalues (percent explained variance) are 5.08 (31.7%), 2.12 (13.5%), 1.15 (7.2%), and 1.04 (6.5%).
[b]Item reverse coded.

The first factor, "no consideration by management" has a high reliability (.86) and high average inter-item correlations (.51). "Ideas not listened to" loads at .50 on factor 3 but has been included in Factor 1 as it fits conceptually with the other items. The scale is a measure of employees' perceptions of management's attitudes toward them. These are attitudes held by management as a result of its position of control. The scale measures the degree to which management expresses a lack of consideration toward workers. "Lack of appreciation by the (immediate) boss," "no appreciation by management," and "not treated with respect by management" deal with not being made to feel needed or treated as a valued and important part of the work organization. "Ideas not listened to" implies not being accepted as making a worthwhile contribution. "Work not appreciated by the supervisor" and "told you are doing a good job" are about being made to feel that one is making a useful and worthwhile contribution to the organization. These attitudes by management can have an important effect on workers' sense of competence and pride in their work. According to Braverman, management has taken the planning role for work while workers simply execute tasks. A lack of consideration is a reflection of management's belief in the right to exercise control over workers. This scale is important in relation to labor process theory.

The second factor, "workload," represents quantitative overload. Research by Caplan and coworkers (83), Karasek (65, 66), and others shows quantitative overload to be an important predictor of stress. Workload represents several aspects of quantitative overload. "More than one task at a time" is a measure of too many tasks at once, "too much work in a limited time" represents a general overload; and "taking work home" is the frequency with which work needs to be done out of regular work hours. Workload is a measure of job demand. The requirement to complete more work than resources allow represents an aspect of management control of the work process and the allocation of resources. As management gains greater control over work processes, especially through the use of mechanized systems and new technological processes, its ability to control job demands increases. Management determines job requirements and the output requirements for work processes and for individual workers. Scale reliability (.64) is high and average inter-item correlation is moderate to high (.38). Factors all load highly on the second factor indicating internal consistency in the scale.

The third factor, "skill discretion," represents the opportunity to use skills and to perform varied tasks on the job. The opportunity to exercise skill discretion over the job represents outcomes of deskilling and job fragmentation. It incorporates combined measures used by Karasek (65, 66): it measures the effects of lacking skill discretion as part of a wide measure of decision latitude. Scale reliability is moderate (.56), as is the average inter-item correlation (.30). "No use made of training" represents the degree to which previously learned skills are used, while "no new skills learned" refers to training in new techniques offered on the job. It is conceptually related to Braverman's idea of deskilling: the opportunity to use skills on the job gives workers a certain amount of control over point-of-production activities. "Lack of job variety" refers to the scope and range of tasks on the job. It loads at .43 on factor 1, but fits conceptually in factor 3 as it relates to Braverman's concept of job degradation through increased specialization and division of labor.

As shown by a number of labor process writers such as Braverman (2), Hill (244), and Thompson (238), skill content and the scope of tasks have diminished as a result of management control of the labor process. Identification with a skills-based craft or a professional or semi-professional group can help workers develop crafts and professional identification. This can increase their ability to rely on knowledge so as to exercise influence over how work is done and can provide a bargaining base with management. Workers performing jobs with no skills base or variety of tasks are easily replaced either by other workers or by increased mechanization.

"Poor relationships," the fourth factor, largely results from the organization of work processes and the way that work is performed. When work organization is centered around mechanized processes, a lot of pressure is placed on personal interaction. Many work processes severely restrict communication. Work over-demand also inhibits satisfying communication. In fact, in some work organizations no talking is allowed on factory assembly lines as it is claimed to slow production rates. The system of work organization instituted by management and the associated rules for interpersonal communication are a way in which management can exert direct forms of control over a workforce. Production processes at the factory level often have associated close supervision, which affects relationships in the workgroup. "Not treated with real feelings" represents being treated with little value, while experiencing "tensions at work" indicates the degree of pressure that is placed on interpersonal relationships. Both items load moderately to highly on factor 3. Even though "tensions at work" loads moderately on factor 1, the scale has proven useful and certainly is a better predictor than either item individually. Scale reliability is moderate (.49), as is the average inter-item correlation (.35).

Finally, "role conflict" and a "lack of freedom" to control the job are single-item measures. "Role conflict" refers to conflicting expectations about how work needs to be done. It represents an aspect of direct control by management via reporting relationships established through the management hierarchy. "Lack of freedom" to exercise control over the job refers to the amount of perceived discretion to exercise control. Management exercises a great degree of control over point-of-production activities, both through direct supervision and through constraints imposed by mechanized processes introduced by management.

A more complex measure of negative work experience has been developed in order to simplify the measurement of effects of negative work on dependent variables (stress, self-esteem, mastery, ill-health, and Goldberg's GHQ). A second-order factor analysis, a factor analysis of factor scales, was undertaken.

All factors from the previous two factor analyses were included, making a total of six factors. Two second-order factors emerged, while an individual factor formed a third dimension on its own (Table 3). The two composite scales were the perception of "influence and control over the task (job) level" together with "negative management attitudes toward workers," and "management control of how work is organized."

The first dimension includes "lack of skill discretion," "no freedom to control work," and "lack of consideration." "Lack of consideration" loads moderately on factor 2 but has been retained in the dimension as a measure of the perception of work experience that includes organizational climate. It fits here conceptually in that the two other factors are measures of perceived decision latitude from a position of low influence: this item reflects management attitudes that are derived from its position of control and directly impinge on work experience. Even though the final two items weigh fairly highly on factor 2, they are related conceptually most strongly on factor 1. Reliability is high (.74), and the average inter-item correlation is moderate to high (.49).

The second dimension, "management control of how work is organized" has a moderate to high reliability (.55) and a low to moderate average inter-item correlation (.38). The factors "role conflict" and "poor relationships" are aspects

Table 3

Rotated factor loadings, reliabilities, and average inter-item correlations: dimensions of negative work experience

		Factor loadings[a]		Scale Reliability	Average inter-item correlation
		I	II		
1.	Influence over task and negative attitudes by management				
	i. Lack of skill discretion	.80	.02		
	ii. Lack of consideration	.61	.47		
	iii. No freedom to control work	.61	.40	.74	.49
2.	Management control of how work is organized				
	i. Role conflict	−.20	.87		
	ii. Poor relationships	.15	.68	.55	.38
3.	Workload	n.a.	n.a.	n.a.	n.a.

[a]Factor loadings were obtained using oblique rotation with unities in the main diagonal. Eigenvalues (percent explained variance) are 2.46 (41%) and 1.44 (24%).

of management's control of work process and structure and are outcomes of increased control. Finally, "workload" is a factor that represents demands in relation to the suitable availability of resources for workers.

The effects of "a perceived lack of control" and "job routine" on occupational stress were included in order to test the effects of these two factors as found in studies by Schwalbe and Staples (7) and others. Schwalbe and Staples used these two measures as important effects of the organization of the labor process on the nature of work for lower level employees. These two elements of work measure the effects of management control at the point of production (task). Other elements of management control dealing with control of structure and climate have been found to have a stronger effect than task-related negative factors. A number of task-related items used in studies by Schwalbe and Staples (7), Karasek (65, 66), and Pearlin and coworkers (178) were included. They are composed of a number of measures of decision latitude used in Karasek's decision latitude scale.

We devised scales through the method of factor analysis rather than using previous scales for several reasons. Other scales do not include the range and composition of control items suitable to a more in-depth study in a manufacturing environment. Even though several validated studies exist, a greater number and variety of measurement devices can enhance our understanding of the causes and consequences of occupational stress. A single-item measure for "job routine" was used as it measures important aspects of negative work experience but has no other items clustered with it.

The factor analysis is shown in Table 4. It includes only items related to task aspects of work. The factor "no control" has three items all representing the perception of influence and control at a number of different levels at work. "Freedom to organize and control work" represents the opportunity to use discretion in performing the job—that is, doing work in the way the worker sees fit. "Ideas listened to" measures the amount of influence over work as perceived by workers. Finally, "influence over policy" is a more general measure of perceived influence over work planning and the way that work is carried out. Both scale reliability (Cronbach's alpha .75) and average inter-item correlation (.50) are high. This is a strong and internally consistent scale. Only one factor emerged from factor analysis, therefore no rotated solution was performed. "Routinization" was retained as a single-item measure of job routine.

Occupational Stress

Job stress has been measured in a variety of different ways in previous studies, as have other dependent variables in the study. These provide an outline of the concept of work stress. It is treated as a state of arousal, a negative emotional state occurring as a response to certain noxious work factors, among other things. It occurs when both personal and social resources are less than sufficient in meeting negative demands or conditions in the work environment. Not all studies

Table 4

Rotated factor loadings, reliabilities, and average inter-item correlations:
the perception of control and influence over work

	Factor loadings[a]	Scale Reliability	Average inter-item correlation
	I		
1. No control			
i. Ideas listened to[b]	.85		
ii. Freedom to organize and control work[b]	.79		
iii. Influence over policy[b]	.73	.75	.50
2. Routinization			
i. Task variety[b]	n.a.	n.a.	n.a.

[a]Eigenvalue (percent explained variance) is 2.47 (61.8%).
[b]Item reverse coded.

of stress use a measure of stressfulness. A measure of emotional reaction to noxious work conditions builds a vital step in the work–stress–ill-health causal relationship, and without it some important characteristics of alienating work environments would be less well known.

A measure of emotional response to negative work conditions (such as degree of bother, worry, or stress) indicates the extent to which certain important needs are not being met. Trouble, worry, or bother experienced shows that the resources available to the worker are not adequate for dealing with demands made. For example, factory workers in the sample indicated they had little opportunity in their work to exercise control, yet they reported that this is not a particularly strong cause of trouble, bother, worry, or stress. However, there is a much higher correlation between the level of control reported by white-collar workers and their level of trouble, worry, or bother. In this way a measure of emotional response gives some understanding of the level of alienation experienced and can offer insights into the sociopsychological process involved. The factory workers, for instance, are likely to have experienced a considerable lack of control in their social life and previous work experience and, as a result, are likely to have learned not to expect to satisfy such a need at work.

There are three types of approaches to dealing with stress in studies of occupational stress. Some studies imply a causal relationship between certain noxious conditions and ill-health outcomes but do not employ a measure of stressfulness (that is, negative emotional response). Other studies measure negative emotional response as an intervening variable between factors such as job strain and somatic and/or psychological outcomes. Yet other studies mainly employ objective indices of stress response, but this does not preclude their using subjective measures of emotional response as well. These different methods of conducting stress research are nonetheless all measuring important aspects of the stress response.

Studies using the first approach do not include a recorded response of stressfulness for negative work factors. In their study of the stress process, Pearlin and associates (178) record a stress outcome response (depression) as a result of job strains. This approach measures a causal outcome of ill-health resulting from job strains and other negative features at work: it implies a causal connection presupposing a stress response which has been well documented in previous research. Karasek (65, 66) and others employ a similar method. Many other studies (e.g., 83, 101) use various scales that record negative emotional responses to job strains and other noxious work conditions. These studies report the degree to which distress (a bother or trouble) is caused by various work conditions. Another quite separate group of studies employs objective indices of stress. Frankenhaeuser (10) records objective measures of the stress response through analyses of adrenaline, noradrenaline, and cortisol in urine and blood samples.

This study uses self-perception as a measure of stress, rather than physiological testing. In Australia, research by the Brain and Behavior Research Institute at La Trobe University has used objective measures of the stress response. Certain disadvantages with this method influenced me to select the method used in my study. Research that involves taking urine samples in the workplace is difficult to organize in terms of gaining cooperation both from a company and from workers, and is expensive to conduct. Finding a suitable site and the required resources would therefore be difficult and costly. I had originally investigated the cost of using physiological measures. The time required for employees from a suitable research setting to participate would have been difficult to organize, and this might have restricted my research sample size considerably. Most importantly, however, self-perceived stress—the measure employed in this study—is a useful measure of the experience of stress: it is an important determinant of how people will act.

People's perception of their situation is very important in understanding stress. Thomas (221), Cooley (222), and others point out that people interpret and thereby give meaning to situations. Stress is a sociopsychological response that leads to a physiological reaction; to measure stress as an outcome of a person's perception of stressfulness is to account for the fact that different people perceive the same situation differently, depending on their socialization and the needs they wish to satisfy. Frankenhaeuser also argues that people's response to stress depends on their own interpretation of the situation: physiological measures of stress still need to be accompanied by a self-report (see 10). High adrenaline levels, for example, may indicate either a positive feeling of high excitement or a negative, uncomfortable feeling of overexertion. Respondents were asked in

Frankenhaeuser's study to indicate whether the feelings that accompanied their adrenaline levels were positive or negative. When accounting for the role of perception in the stress response, an event may be perceived as threatening to one person and not to another. This may lead to stress for some people, whereas for others there may be few grounds for considering the event a stressor.

There are many comparable studies using self-perception measures to permit adequate comparison. Self-reports represent a fruitful indication of stress outcomes: sociopsychological arguments on the nature of the stress response show that individuals' evaluations of their own level of negative emotional response are a pertinent indication of how effectively their personal resources are able to deal with inordinate demands from the environment.

The stress scale measures the degree to which people perceive they are stressed by each of the work items listed in Table 1, by asking "over the past 12 months has this aspect of your job bothered, worried, or stressed you?" All items are related to outcomes of management control. Respondents were asked to indicate their response on a scale of 0–3: 0, not at all; 1, a little; 2, a fair bit or moderately often; and 3, a lot or very much.

Self-Esteem Scale

A small number of self-esteem scales are available for use in occupational research. The Coopersmith Self-Esteem Inventory Adult form composed of 25 items is one example. I chose to use the Rosenberg Self-Esteem Scale for a number of reasons. It is a short, simple scale taking very little time to complete. It asks very straightforward questions, compared with other scales such as the Coopersmith (281), encouraging cooperation from survey participants. Finally, it is a unidimensional scale that seeks to measure the one state of self-worth.

Wylie argues that "Rosenberg [257] is apparently the only person who has tried to achieve a unidimensional measure of global self regard (called by him "Self-esteem") through the method of Guttman scaling" (282, p. 181). Rosenberg maintains that his scale is unidimensional; according to Wylie (282), unidimensionality is well accepted. Rosenberg argues for the scale that (257, p. 16):

> an instrument was required which would enable us to rank people along a single continuum ranging from those who had very high to those who had very low self-esteem. The Guttman scale ensures a unidimensional continuum by establishing a pattern which must be satisfied before the scale can be accepted.

He maintains that the reproducibility of the scale is 92% and scalability is 72%. Rosenberg argues that the terms established by Guttman and Menzel show coefficients to be satisfactory. Silber and Tippett (283) report a test reliability coefficient of .85.

Table 5

Rosenberg's Self-Esteem Scale: mean scores

		Mean	(S.D.)
1.	I feel that I am a person of worth, at least on an equal basis with others.	.46	(.55)
2.	I feel I have a number of good qualities.	.49	(.53)
*3.	All in all, I'm inclined to feel that I am a failure.	.49	(.62)
4.	I am able to do things as well as most other people.	.60	(.65)
*5.	I feel that I do not have much to be proud of.	.53	(.69)
6.	I take a positive attitude toward myself.	.75	(.68)
7.	On the whole, I am satisfied with myself.	.92	(.66)
*8.	I wish I could have more respect for myself.	.95	(.83)
*9.	I certainly feel useless at times.	1.00	(.83)
10.	At times I think I am no good at all.	.63	(.78)

*Item reverse coded before scaling.

Rosenberg's Self-Esteem Scale (10 items) is included as a measure of outcomes of work experience. The scale is scored from 0 to 3: employees were asked whether they, 0, strongly agree; 1, agree; 2, disagree; or 3, strongly disagree with the statements listed in Table 5.

Sense of Personal Mastery Scale

Various component scales are available to measure internal locus of control. However, many of these scales also incorporate components of personality in their measurement. The test in this model required simply mastery (internal locus of control). Pearlin's sense of mastery scale, which uses seven questions to test for sense of control over external events, is appropriate to an adult population in an occupation setting. It also allowed some direct comparison's with Pearlin and Schooler's (99) and Pearlin and associates's (178) research and with research using modified versions of that scale.

Pearlin and Schooler (99) do not report the factor structure for the scale. As used by Pearlin, the scale is treated as a unidimensional measure. For items in the scale, responses were scored from 0 to 3. Respondents were asked to indicate whether they, 0, strongly agree; 1, agree; 2, disagree; or 3, strongly disagree with the seven statements listed in Table 6.

The scale was factor analyzed and two factors emerged (Table 7). The first contained items relating to perceived decision-making and problem-solving effectiveness, and the second to more general control of future possibilities. Neither of these factors is used as a single measure in the study. The first factor, "perceived decision-making and problem-solving effectiveness," contains four

Table 6

Sense of mastery scale: mean scores

		Mean	(S.D.)
*1.	There is really no way I can solve some of the problems I have.	.91	(.75)
*2.	Sometimes I feel that I am being pushed around in life.	1.19	(.83)
*3.	I have little control over the things that happen to me.	.80	(.65)
4.	I can do just about anything I really set my mind to.	.84	(.66)
*5.	I often feel helpless in dealing with the problems of life.	.89	(.67)
6.	What happens to me in the future mostly depends on me.	.66	(.67)
*7.	There is little I can do to change many of the importnat things in my life.	.86	(.64)

*Item reverse coded before scaling.

items. Reliability as measured by Cronbach's alpha is high (.80), as is the average inter-item correlation (.50). This factor indicates a lack of ability to change current situations. The second factor, "control over future states," has a moderate to high reliability (.61) and moderate inter-item correlation (.34). It is comprised of three items. This factor also indicates a lack of ability to change current situations. The third item weighted .52 on factor 1 but fits conceptually in factor 2. The reliability score for the whole scale is .80; the average inter-item correlation is moderate to high (.37). Israel and coworkers (101) employ a mastery scale composed of only three of the seven items used by Pearlin and Schooler (99). This need not be problematic for comparative purposes if similar factor structures emerge; evidence was not available from Israel and associates on factor structure, reliability, or validity.

Ill-Health Symptoms Scale

Various measures of somatic and psychological symptoms have been employed in studies of occupational stress outcome. A scale measuring self-perception of symptoms is a valuable way of ascertaining somatic and psychological ill-health symptoms. The scale used in this study is a measure of symptom awareness.

Mechanic (63) argues that "individuals' awareness of morbidity" depends on their own knowledge, and underreporting can be a problem. A report of symptom awareness does not necessarily equate with symptom diagnosis. Mechanic (63) reports that while women record a greater symptom awareness than men, their diagnosis level of symptoms has been found to be similar. However, Dorn (284) argues that the self-perception of illness is strong evidence of ill-health despite the presence or absence of measurable symptoms. Broadhead (285) reports that Anderson (286) shows a high correlation between self-reports of physical conditions and clinical tests in that, if symptoms are perceived, they are a real barrier to

Table 7

Rotated factor loadings, reliabilities, and average inter-item correlations:
mastery or personal control

	Factor loadings[a]		Scale Reliability	Average inter-item correlation
	I	II		
1. Decision-making and problem-solving				
i. I have little control over the things that happen to me.	.80	.21		
ii. Sometimes I feel I am being pushed around in life.	.79	.06		
iii. I often feel helpless in dealing with the problems in life.	.75	.23		
iv. There is really no way I can solve some of the problems I have.	.73	.18	.80	.50
2. Control over future states				
i. What happens to me in the future depends mainly on me.[b]	.13	.78		
ii. I can do just about anything I set my mind to.[b]	.13	.75		
iii. There is little I can do to change many of the important things in my life.	.52	.55	.61	.34

[a]Factor loadings were obtained using Varimax rotation with unities in the main diagonal. Eigenvalues (percent explained variance) are 3.26 (46.5%) and 1.01 (14.4%).
[b]Item reverse coded.

social functioning. The value of a self-perception symptom index is backed by Broadhead (285): for the sociologist, self-perceived symptoms are indications of an effect on normal functioning and role behavior. This evidences a mediating role played by social factors in the consequences of perceived illness. The perception of symptoms can lead to real consequences in terms of social function.

This research used a symptom awareness checklist for two reasons. First, such checklists are used extensively in similar research into the relationship between stress and ill-health outcomes, and second, the cost involved in clinically testing the survey population and the availability of a suitable site were prohibitive. No single standardized symptom list for measuring stress outcomes is used across stress studies. Caplan and coworkers (83), Pearlin and associates (178), Schwalbe and Staples (7), and Israel and associates (101), among others, have employed different measures of somatic and psychological ill-health. Pearlin and coworkers (178) and Karasek (66) employed a scale of depression as a measure of ill-health outcomes of stress. Caplan and coworkers (83) measured emotional symptoms, among other measures, for a wide range of occupational groups. Schwalbe and Staples (7) used a scale of a small number of somatic and emotional items for self-report.

This study employed a scale of 27 physical and emotional symptoms developed from scales used in a variety of previous studies, including those of Caplan and associates, Schwalbe and Staples, and Pearlin and associates. Two of the 27 items were not included in the scale due to a large number of missing responses: these were "trouble with menstrual periods" and "do menstrual periods cause any problems for you at work?" Respondents were asked to record on a scale of 0–2 if, in the past 12 months, they had experienced any of these symptoms: 0, hardly ever or never; 1, sometimes; or 2, often. The 25 items in the symptom scale are shown in Table 8.

A factor analysis was carried out of all items in the 25-item symptom aware-ness scale (Table 9). Even though "urinary problems" is a conventional symptom of anxiety, an emotionally related disorder (287), it is not included in the first factor. Its loading on factor 1 was low and it loaded equally (.38) on factor 3. Two other items have not been included in the group of factors: "pains in the heart, tightness or heaviness in the chest" and "unusual heartbeats."

The first factor, "emotional symptoms," has a high reliability (.84) as measured by Cronbach's alpha and a high average inter-item correlation (.43). All items have high loadings on the first factor and low loadings on all other factors. Seven symptoms are included; these are reported by Braunwald (287) to be conven-tional symptoms of emotional complaints. Reduced interest in sex is a likely consequence of fatigue or depressive-related conditions. Emotional and mental complaints include depression as well as anxiety symptoms. Various studies have used components of these measures or separate elements. Lowe and Northcott (146) use measures for depression, anxiety, and psychophysiological symptoms. Caplan and coworkers (83) also employ separate scales for depression, anxiety, and irritation. Karasek (65, 66) Pearlin and associates (178) use measures of depression symptoms.

Other studies (e.g., 83) include feelings of sadness/happiness and feeling good/blue/cheerful as measures of depression. Cassem (288) classifies fatigue and listlessness as an emotional symptom, along with depression, nervousness, and anxiety. These are likely outcomes of work experience that creates pressure and frustration. Frankenhaeuser (10) reported that machine-paced work and shiftwork lead to increased adrenaline levels, which can have emotional conse-quences; other studies show that overdemand and underdemand can lead to emotional fatigue or depression. Tearfulness is also an indication of a strained emotional state. Caplan and associates' (83) study includes nervousness as one of

Table 8

Frequency of ill-health symptoms: mean scores

		Mean	(S.D.)
1.	Headaches	.67	(.70)
2.	Stomach pains	.29	(.51)
3.	Loss of appetite	.24	(.51)
4.	Spells of dizziness (or fainting)	.16	(.39)
5.	Nervousness, tenseness	.85	(.69)
6.	Feeling of depression	.71	(.69)
7.	Difficulty with falling asleep or staying asleep	.72	(.75)
8.	Getting tired easily or very quickly	.88	(.73)
9.	Sore throats or colds	.52	(.63)
10.	Trouble breathing or shortness of breath	.35	(.63)
11.	Constipation	.22	(.46)
12.	Diarrhea	.16	(.41)
13.	Pains in the heart, tightness or heaviness in the chest	.29	(.56)
14.	Nausea or vomiting	.09	(.33)
15.	Swelling or aching in any joint or muscle	.56	(.70)
16.	Pains in the back or spine	.69	(.72)
17.	Pains in the neck or arms	.63	(.74)
18.	Pains in the lower back or legs	.61	(.70)
19.	Unusual heartbeats	.13	(.35)
20.	Need to pass urine frequently	.36	(.61)
21.	Skin trouble	.33	(.62)
22.	Irritability	.74	(.67)
23.	Tearfulness	.31	(.57)
24.	Reduced interest in sex	.48	(.63)
25.	Feeling generally run down, low in energy	.88	(.70)

four measures of anxiety and irritation. The scale is a broad range of depressive/anxiety-related complaints.

The second factor, "pains in the back and neck," includes a number of spinal-related complaints that are typical musculoskeletal conditions. Lower back pains and back pains can be related to postural discomfort, incorrect lifting, and other movement. Mankin and Adams (289) included aching joints and neck pains as a general group of spinal-related symptoms; these can be related to some occupational overuse syndrome (OOS) symptoms. However, the measure as used here is intended as a broader measure of outcomes of the work experience including accident, postural discomfort, and general working conditions. Reliability is high (.79), as is the average inter-item correlation (.48).

"Gastrointestinal symptoms," the third factor, has a moderate to high reliability (.67) and a high average inter-item correlation (.40). The three items—nausea,

Table 9

Rotated factor loadings, reliabilities, and average inter-item correlations:
symptom awareness

	Factor loadings[a]						Scale Reliability	Average inter-item correlation
	I	II	III	IV	V	VI		
1. Emotional symptoms								
Reduced interest in sex	.71	.12	.01	−.11	−.02	−.07		
Low in energy	.67	.21	.16	.17	.22	.03		
Feelings of depression	.67	.09	.04	.09	−.06	.33		
Irritability	.63	.08	.22	.23	.18	.20		
Nervousness	.60	.14	.23	.07	.16	.31		
Easily tired	.57	.18	.20	.30	.13	.04		
Tearfulness	.57	.27	.10	.13	.30	.04	.84	.43
2. Pains in the back and neck								
Lower back pains	.10	.82	.07	.12	−.02	.11		
Back pains	.13	.77	.15	.04	.10	.06		
Aching joints	.27	.71	.05	.11	.02	−.11		
Neck pains	.37	.58	.10	.11	.32	.14	.79	.48
3. Gastrointestinal symptoms								
Nauseau	.13	.03	.70	.14	.05	.04		
Diarrhea	.20	.05	.68	−.11	−.06	.17		
Stomach pains	.01	.38	.68	.05	.16	.19	.67	.40
4. Allergic symptoms								
Sore throats or colds	.03	.15	.12	.77	.15	.14		
Skin trouble	.25	.12	.23	.62	−.03	−.15		
Breathing trouble	.45	.11	.19	.47	−.16	.21	.52	.27
5. Analgesic response								
Constipation	.14	.06	−.10	−.09	.76	.04		
Headaches	.16	.16	.31	.23	.58	.19		
Dizziness	.27	.01	.43	.12	.45	−.14	.47	.23
6. Functional depressive disorders								
Loss of appetite	.10	−.03	.38	.12	−.03	.72		
Sleep difficulties	.38	.37	−.03	−.02	.18	.64	.52	.35

[a]Factor loadings were obtained using Varimax rotation with unities in the main diagonal. Eigenvalues (percent explained variance) are 6.87 (29.9%), 1.65 (7.2%), 1.50 (6.5%), 1.24 (5.4%), 1.11 (4.8%), and 1.05 (4.5%).

diarrhea, and stomach pains—load highly only on the third factor, and low on other factors. Isselbacher and May (290, p. 1223) suggest that gastrointestinal symptoms occur not only with primary gastrointestinal tract disease but frequently as manifestations of other organ and functional disorders. Peptic ulcers are also suggested. Nausea and diarrhea are commonly associated symptoms of this type of disease, as are pains in the stomach or abdomen. Infective disorders are a likely condition with this set of symptoms.

The fourth factor, "allergic symptoms," is characterized by sore throats or colds, skin trouble, and breathing problems: these are suggestive of allergic rhinitis (291). These types of ailments are likely to occur in work situations with dirt or dust in a factory setting or in an air-conditioned environment with recirculated air. Austen (291) suggests that dust can have a mixed content, including mites; this can be a cause of allergic symptoms. He argues that hay fever is not an appropriate term for the condition. The reliability score is moderate (.52), and the average inter-item correlation is quite low to moderate (.27).

The fifth factor, "analgesic response," is another functional emotional disorder that is likely to be associated with analgesic overuse. The effects of overuse of codeine are an example of drug use that can manifest in this group of conditions. Goodman and Gilman note that narcotic analgesics such as codeine may have "a wide spectrum of unwanted effects" including dizziness and constipation, and go on to argue that "these occur so commonly that they cannot be considered idiosyncratic" (292, p. 257). Headache, according to Goodman and Gilman, is a symptom of withdrawal from codeine: this is a similar effect to that seen in withdrawal from methadone or morphine. The symptoms in this group are also likely to be functional emotional symptoms, which typically can be exacerbated by the overuse of analgesics such as codeine. While the alpha reliability (.47) and the average inter-item correlation (.23) are low, they are not too low to be of value for certain analyses. This also represents a definite symptom group.

Finally, appetite problems and difficulty with sleeping are two symptoms associated with endogenous depressive states. They represent more severe and somatic symptoms than the items included in the emotional symptoms scale. Cronbach's alpha reliability score is moderate (.52), as is the average inter-item correlation (.35).

Psychiatric Impairment:
Goldberg's General Health Questionnaire

Goldberg's GHQ (12-item version) has been included as a measurement tool of psychiatric impairment. It is a well-validated and reliable standard health index which is used as a measure of stress outcome. There are few scales comparable to Goldberg's GHQ. Most scales that measure psychiatric impairment also measure a personality dimension; the GHQ does not. The SCL-90 checklist (293) is a frequently used symptoms test, but it also includes personality dimensions and was therefore excluded from use in this study.

Clinical tests to measure levels of ill-health have been used extensively in health surveys. Goldberg's GHQ has been employed in a number of studies, including the Australian Health Survey. There are various versions of the GHQ, the most common having 60, 30, 28, and 12 scale items. The GHQ, according to Goldberg (294, p. 5):

> was designed to be a self-administered screening test aimed at detecting psychiatric disorders among respondents in community settings, such as primary care among general medical outpatients. . . . It is aimed at detecting those forms of psychiatric disorder which may have relevance to a patient's presence in a medical clinic.

The GHQ aims to evaluate nonpsychotic mental disturbance, and it has been validated widely in the United States, Great Britain, and Australia (see 279, 295, 296). In Australia, Tennant (295) validated use of the GHQ when 120 patients at two general practice clinics in Sydney completed interviews and the GHQ. The 60-item scale was used, and scores were calculated for the 12-item and other versions of the GHQ. Tennant (295, p. 393) reports that the validity of the findings was very similar to that obtained by Goldberg in his sample of general practice patients in England. He, along with others, found the GHQ to be an effective measurement tool. Tennant found it "an efficient, reliable and valid index of non-psychotic psychological impairment" (295, p. 392).

While it was originally developed for use in general practice situations, the GHQ has been validated for use in other settings (e.g., 297). The 60- and 30-item versions have been factor analyzed (see 298–300). The first reported factor analysis of the 12-item GHQ using an Australian sample population (603 householders) was by Worsley and Gribbin (301). They report three orthogonal factors: sleep disturbance, diminished social performance, and loss of confidence; they provide evidence of factor stability across samples and between other versions of the GHQ. Factors 2 and 3 show that the scale measures factors other than psychiatric disturbance.

In Goldberg's GHQ (the 12-item version), respondents are asked to indicate how their health has been, in general, over the past few weeks. The scale records responses on a 0–3 scale: 0, better than usual, much more than usual, or more so than usual; 1, same as usual, rather more than usual, or about same as usual; 2, less than usual, no more than usual, or less so than usual; and 3, much less than usual, not at all, or much less than usual. The 12 scale items are shown in Table 10.

A factor analysis was carried out using a rotated Varimax solution. Two factors emerged (Table 11). The first factor "anxiety and personal disturbance," identifies principally psychiatric disturbances relating to personal feelings of pleasure and

Table 10

General Health Questionnaire: mean scores

		Mean	(S.D.)
1.	Been able to concentrate on whatever you're doing?	1.20	(.53)
*2.	Lost much sleep over worry?	.90	(.77)
3.	Felt that you are playing a useful part in things?	.92	(.61)
4.	Felt capable of making decisions about things?	.91	(.54)
*5.	Felt constantly under strain?	1.12	(.78)
*6.	Felt you couldn't overcome your difficulties?	.76	(.78)
7.	Been able to enjoy your normal day-to-day activities?	1.17	(.59)
8.	Been able to face up to your problems?	1.02	(.51)
*9.	Been feeling unhappy and depressed?	.96	(.85)
*10.	Been losing confidence in yourself?	.68	(.81)
11.	Been thinking of yourself as a worthwhile person?	1.96	(.51)
12.	Been feeling reasonably happy, all things considered?	1.08	(.60)

*Item reverse coded before scaling.

adequacy. The second factor, inadequate social performance," grouped items less related to psychiatric disturbance. These were assessments of capable performance and decision-making characteristics. The two factors, rather than the three found by Worsley and Gribbin (301), may have resulted from characteristics of the sample size and composition. Factor 1 has a scale reliability of .84 and an average inter-item correlation of .40. The reliability for factor 2 is smaller (.58), and the average inter-item correlation is small (.26).

The GHQ is a valid and reliable measure of nonpyschotic psychiatric symptoms that has a tested relevance outside clinical settings. As such, it is a suitable tool for measuring ill-health along with the Symptoms Inventory (a measure of perceived physical and emotional symptoms). It is relatively free of influence from factors such as age, gender, and racial origin. Scores on the measure are not directly affected by physical health symptoms. Thus it provides a health measure discretely different from the measurement of physical health symptoms. The GHQ is a reliable and valid test of psychiatric disturbance. Although the 12-item version of the GHQ is treated in this study as a single scale, Graetz argues that due to the clinical and diagnostic nature of the GHQ, "it is reasonable to suppose that the GHQ detects various dimensions of symptomatology" (296, p. 132). He argues for an analysis of factor structure based on oblique rotation, one that is a simpler solution subject to easier interpretation, rather than on the orthogonal method of factor analysis used by Worsley and Gribbin (301). Using oblique rotation, Graetz identifies the separate factors as "anxiety," "social dysfunction," and "loss of confidence."

Table 11

Rotated factor loadings, reliabilities, and average inter-item correlations:
General Health Questionnaire (GHQ)

	Factor loadings[a]		Scale Reliability	Average inter-item correlation
	I	II		
1. Anxiety and personal disturbance				
i. Feeling unhappy and depressed	.77	.18		
ii. Constantly under strain	.74	.01		
iii. Lost sleep over worry	.73	−.25		
iv. Couldn't overcome difficulties	.68	.35		
v. Losing confidence	.62	.43		
vi. Enjoy normal day-to-day activities[b]	.59	.36		
vii. Reasonably happy all things considered[b]	.57	.36		
viii. Concentrate on whatever you are doing[b]	.53	.18	.84	.40
2. Inadequate social performance				
i. Capable of making decisions[b]	.23	.67		
ii. Thinking of yourself as a worthwhile person.	−.15	.62		
iii. Playing a useful part in things[b]	.15	.59		
iv. Face up to problems[b]	.39	.53	.58	.26

[a]Factor loadings were obtained using Varimax rotation with unities in the main diagonal. Eigenvalues (percent explained variance) are 4.50 (37.5%) and 1.37 (11.4%).
[b]Item reverse coded.

CONCLUSION

As is evident from the literature review in earlier chapters, a great deal of research has been conducted into the causes of occupational stress and ill-health. The causes of occupational stress are not always presented in a clear framework, however. The nature of the problem of occupational stress is most clearly presented by labor process writers: differences in occupational stress and ill-health are best understood as outcomes of management's attempts to exert control over the labor process. Alienating work that leads to stress and possibly ill-health results most often from differences in control at work.

The questionnaire method allows for the testing of factors contributing in a causal sequence to stress and ill-health. The choice of a manufacturing location was shown to provide a relatively consistent environment within which occupational and gender experiences could be compared and contrasted. A large number of quantitative studies have tested the influence of causal factors on stress, but few have employed a comprehensive sociostructural framework. Few have shown differences in occupational and gender groups, and still fewer have examined these within a more in-depth and integrated research environment.

The causal model of all the factors tested in the sample forms the basis of all tests carried out in the following five chapters. I will refer to the model presented in this chapter in the presentation of results.

Each of the major measures—alienating work, occupational stress, low self-esteem and low sense of personal mastery, somatic and emotional ill-health, and psychiatric impairment (GHQ)—were selected as reliable measures suitable for occupational settings. Where no suitable measures existed, scales were constructed (alienating work, occupational stress, and ill-health). All items from the scales that are used in the analysis have been presented with mean (average) scores for the total sample together with standard deviations, in order to show, prior to the analysis, the total components of the survey data. Composite scales will be used for further analyses based on factor analyses of the negative work scale and the physical and emotional symptoms scale.

The causal model presented in this chapter is the means through which the sociological problem—the differential experience of stress between occupational and gender groups—is addressed. All tests in the following five chapters are intended to test the nature and extent of sociostructural relationships in this causal model.

CHAPTER 7

Management Control and Its Effects on Alienating Work Experience

The research reported in this book is based on the working lives of people employed at various occupational levels in one large manufacturing company. It investigates structural relationships that lead to stress in this particular organization. This chapter presents a brief history of technological and other related developments in the industry and company, and describes the work environment of the company in order to highlight the context within which stress and ill-health take place. The remaining chapters of the book present the tests for the effects of occupational position and gender on several dependent variables, including negative work factors, stress, self-esteem, sense of mastery, and associated health outcomes.

In this chapter I deal with a number of issues. After giving an overview of industry characteristics and details of the sample, I undertake a preliminary analysis of achieved and ascribed status characteristics in the study. I then examine the determinants of negative work experience and describe occupational and gender differences in alienating work characteristics.

OVERVIEW OF THE INDUSTRY

In manufacturing, the introduction of complex mechanized processes and advanced technology has brought about substantial increases in management's control of the labor process over the past few decades. It has brought about deskilling in some jobs and a shift in employment opportunities for skilled craft labor.

The company in the study is in the food, beverages, tobacco, and allied industries. It has one manufacturing plant in Australia, located in Melbourne. The company employs more than 1,300 people nationally, with an annual sales of just under $1,000 million. It is part of a multinational organization: the major holding company is located in the United States. It is one of the largest consumer packaged goods companies in the world. Internationally, the company has an operating revenue of over $50,000 million and markets over 3,000 products. The parent company has acquired a number of wholly owned subsidiary companies in its international structure and at the time of the study had five major operating companies.

Ownership of the company, as with many other Australian companies, moved during this century from local to national, and then to multinational. Production processes in the company had been virtually unchanged since the late 19th century. Machine processes were not upgraded until around the 1950s when variable paced machines were introduced. This brought about full mechanization, as machine systems could automatically produce at several times the speed of the older machines because they required less set-up time. However, greater time and motion demands were made on operators, and workloads increased substantially. Reports from the company at the time of the new mechanization showed employees were aware of greater job strains.

Machine process operations require unskilled labor, as has been the case for several decades. Women have been employed as machine operators and as factory floor help during this time.

Large-scale deskilling took place in the manufacturing industry with the introduction of more complex mechanized production processes in the early 20th century and until the 1930s. Changes were seen first in the United States and Britain, but moved rapidly to other industrialized countries. Since then a large pool of "process" or unskilled factory workers (many of them women) has taken on work that had traditionally been performed by skilled craftspeople (see 2). Women more than men have traditionally been employed on the shop floor, due to their lower wage rates. Since the 1950s, during economic downturns married women were the first group to be retrenched.

In the 1960s and 1970s mass skill loss became a concern for middle-class workers, due in part to the introduction of computer technology. Microchip-based robotics became a cause of skill changes. There have been large increases in unemployment in some highly mechanized industries, and despite government and union initiatives, advanced technology continues to affect employment levels. However, new specialist and skilled-based occupations have grown in response to technological development, giving middle-class workers a stronger influence over the work environment. Yet factory work has remained a basic, unskilled activity with an associated lack of fulfillments and rewards for workers.

Multinational considerations largely determine policy creation and implementation in the Australian-based company. While employees on the factory floor have little influence due to the technological demands and constraints of their work, management often has limited corporate influence due to multinational considerations. Management implements policy decisions that often are formulated based on events occurring in the parent organization in the United States.

CHARACTERISTICS OF THE SAMPLE

Table 12 shows the distribution of men and women in the sample, broken first into 11 occupational categories and then classified into five major groups. The major occupational groups are managers, supervisors, and white-collar (office) workers comprising white-collar groups, and skilled and factory workers comprising blue-collar groups. The sample includes 242 employees. Because of missing data on gender, 101 women and 133 men make up the gender proportions of the survey population (not accounting for occupational position). This

Table 12

Sample population by occupational group and gender (number)

	Women	Men	Total	Total employed
Managers				
Senior managers	0	20	20	
Managers	9	27	36	
Total	9	47	56	157
Supervisors				
Clerical	11	8	19	
Sales	1	2	3	
Factory	10	18	28	
Total	22	28	50	144
White-collar (office)				
Clerical	25	5	30	
Sales	6	9	15	
Total	31	14	45	112
Skilled				
Skilled (apprenticed)	1	13	14	
Semi-skilled	3	12	15	
Total	4	25	29	209
Factory				
Machine operators	29	12	41	
Floor help	3	5	8	
Total	32	17	49	533
Total*	**98**	**131**	**229**	**1,155**

*These figures take into account the occupational membership.

occupational categorization was selected over others, such as those discussed by Wright (220), although all supervisors were grouped together as he suggests. Wright suggests that managers should be treated separately from white-collar and blue-collar workers, who should be grouped together. However, few studies have been conducted on differences in stress between blue- and white-collar workers, and this lack of previous research justified a separate analysis of these groups in this study.

The analysis here is of occupational groups and focuses on occupational as well as gender differences. Many of the effects of the different proportional representations of occupational groups are accounted for by this type of analysis. Both white-collar groups and blue-collar groups (skilled and factory) are proportionally represented.

At the time of study, one-third of employees were staff (managers, monthly paid staff, supervisors, salespersons, clerical supports, and quality assurance) and two-thirds were unionized including skilled and semi-skilled tradespersons, machine operators, floor help, drivers, and storepersons).

Managers in the sample were senior, middle, and lower level, performing a variety of functions. Senior management included a board of directors and functional heads such as personnel and finance. Middle management has been a large growth area during the past two decades; these managers performed more specific functions in the company such as training, financial, and budgetary management.

There were several categories of supervisors. Clerical supervisors organized and overviewed, the work of clerical and typing employees; there were also computing operations, accounts, and administrative supervisors. Sales supervisors, located in suburban areas of Melbourne, managed small groups of sales staff including representatives. Factory supervisors, the third major group, managed machine operators and floor help. However, they seldom worked on site. One machine operator described visits by the factory supervisor as resembling a management tour. The supervisors were said to dress like management, even wearing ties. This created some barriers to workers and supervisors communicating effectively.

The white-collar (office) workers were employed mostly on site, although some were located in suburban and rural offices. They included secretaries, word-processing operators, finance and accounting personnel, and computer operators. Sales staff and representatives worked away from the head office. White-collar employees included junior secretarial staff to specialist accountants and computer personnel.

Blue-collar employees included skilled, semi-skilled, and unskilled workers. Skilled apprenticed workers included fitters and turners, electricians, special electricians, electronic tradespersons, carpenters, boiler attendants, and painters. Semi-skilled workers included employees such as forklift operators. The lowest status employees were unskilled factory workers. Attendants and floor help had the lowest status: they sweep and clean around the machinery and carry out related types of work. A floor helper can be promoted to machine operator. Machine operators ensure that machine processes perform effective production tasks. It is a more responsible job than attendant or floor help. These workers can be promoted to leading hand, leading a machine-based workgroup.

PRELIMINARY ANALYSIS
OF BIOGRAPHICAL CHARACTERISTICS

In the causal model presented in Chapter 6 (Figure 1, p. 103), ascribed and achieved variables are the first factors that affect the experience of stress and ill-health. They have direct effects on occupational position and specifically on the experience of negative work. There is conflicting evidence for the role of each status variable as a predictor of work experience, stress, and ill-health. They have been included because previous research has shown them to have some effect, and in some cases (partnership status and number of children) they were found to have some influence in exploratory stages of this research.

The only status variable presented here that is omitted from further analysis is "number of years with the company." It has far less predictive value on major dependent variables than "years in the one position" and other variables. It managed only small to moderate correlations with other variables and, based on other studies of stress, has a less useful theoretical base for inclusion than the other status variables.

Age. Table 13 shows that more women (60%) than men (45%) in the sample were 35 years or younger. Very few people were older than 55 years. There are some differences between white- and blue-collar women in the sample—for example, a greater proportion of younger female managers, supervisors, and white-collar (office) workers than blue-collar women. White-collar men were the youngest in the sample. However, there are few age differences between occupational groups, irrespective of gender.

Education. As can be expected, both female and male managers had the highest proportions of tertiary-educated employees. Table 14 shows the breakdown into educational categories for all groups including managers. Factory women (97%), female supervisors (86%), and factory men (76%) had most employees with only primary or secondary education. A large proportion (76%) of white-collar office women had secondary level education or less. Supervisors were more highly educated than the groups from which they are likely to be recruited.

Number of Children. The average number of children for employees in the sample was 1.0. Men had more children (mean 1.1) than women (0.8) (Table 15). Male supervisors had the largest number of children (mean 1.5), followed by

Table 13

Age categories by gender within occupation (percent in categories)

	25 and under	26–35	36–45	46–55	56 and over	Total (n)	
Women							
Managers	—	62	25	13	—	100	(8)
Supervisors	18	45	23	9	5	100	(22)
White-collar (office)	34	28	24	7	7	100	(29)
Skilled	50	—	50	—	—	100	(4)
Factory	12	19	44	25	—	100	(32)
Total (n)	21 (20)	39 (29)	30 (30)	14 (13)	3 (3)	100	(95)
Men							
Managers	4	39	33	22	2	100	(46)
Supervisors	4	26	52	14	4	100	(27)
White-collar (office)	50	21	29	—	—	100	(14)
Skilled	8	28	36	28		100	(25)
Factory	17	47	12	12	12	100	(17)
Total (n)	12 (15)	33 (43)	34 (44)	18 (23)	3 (4)	100	(129)
Total (n)	**16 (35)**	**32 (72)**	**33 (74)**	**16 (36)**	**3 (7)**	**100**	**(124)**

factory women (1.2). Female white-collar workers (mean 0.5) and supervisors (mean 0.7) had fewest children.

Number of Years in the Same Position. Caplan (83, p. 114) argues that "the person's length of service in his current job . . . [has] potential value as an indicator of length of exposure to stressful and non stressful environments." Length of experience in the job is an important independent variable. In the study sample, a high level of stability is shown by an average of 6.1 years in the same position. This also indicates relatively poor promotion opportunities. Female factory workers had been employed longest in the one position (average 11.2 years), followed by male skilled workers (8.4 years). Women's lack of mobility is due partly to lower levels of education and partly to pressures from their domestic roles inhibiting career moves (302). Managers had spent the shortest time in their current position (3.7 years), partly reflecting their higher level of education (Table 16). White-collar workers had an average of 4.3 years in their current job, with no differences between women and men. White-collar workers could look to management ranks for promotional opportunities, while blue-collar workers were least mobile due to a lower level of education and restricted career paths for promotion.

Table 14

Educational categories by occupation within gender (percent in categories)

	Primary only	Some secondary	Appren-ticeship	Some tertiary	Total (n)
Women					
Managers	—	11	11	78	100 (9)
Supervisors	—	86	—	14	100 (22)
White-collar (office)	6	68	3	23	100 (31)
Skilled	—	75	25	—	100 (4)
Factory	9	88	—	3	100 (32)
Total (n)	25 (5)	74 (72)	3 (3)	18 (18)	100 (98)
Men					
Managers	—	15	19	66	100 (47)
Supervisors	—	53	18	29	100 (28)
White-collar (office)	7	50	7	36	100 (14)
Skilled	—	44	39	17	100 (23)
Factory	—	76	18	6	100 (17)
Total (n)	1 (1)	40 (52)	21 (27)	38 (49)	100 (129)
Total (n)	**3 (6)**	**55 (124)**	**13 (30)**	**29 (67)**	**100 (227)**

Table 15

Number of children by occupational group and gender

Occupational Group	Mean no. of children[a]		
	Women	Men	Total
Managers	—	1.1	1.0
Supervisors	0.7	1.5	1.1
White-collar (office)	0.5	0.9	0.6
Skilled	—	1.0	1.1
Factory	1.2	0.8	1.0
Total	0.8	1.1	1.0

[a]Dashes indicate less than ten cases: insufficient sample size to make an adequate comparison.

Table 16

Number of years in current position by gender

Occupational Group	Mean no. of years		
	Women	Men	Total
Managers	—	3.7	3.4
Supervisors	4.8	6.8	5.9
White-collar (office)	4.3	4.3	4.3
Skilled	—	8.4	7.6
Factory	11.2	8.0	10.1
Total	6.4	5.8	6.1

[a]Dashes indicate less than ten cases: insufficient sample size to make an adequate comparison.

Number of Years with the Firm. Female factory workers had been with the company the longest (average 11.3 years), indicating an environment of stable employment at the time of the study (Table 17). Of the 32 factory women, 27 had working partners. Male managers and supervisors, female white-collar workers, and male skilled and factory workers had been with the company for an average of about nine years. Male white-collar workers had been with the company the least time (4.8 years); they were likely to get relatively quick promotion to supervisory and management positions.

RESULTS

Method of Presenting Results

Results for major hypotheses are presented in the next five chapters. Regression analysis is the main statistical method used.

Tests on expectations about the nature of relationships between negative work, stress, and other sociopsychological states and ill-health have been based on causal relationships found in previous research. However, exploratory research was undertaken in order to test associations between variables where little previous research had been undertaken.

Method of Data Analysis: Multivariate Analysis

The principal method employed is multivariate analysis through the technique of multiple regression. The method of multivariate analysis has a number of advantages over univariate analysis for the type of sample and for the mix of variables and factors in this study. The sample is divergent and contains a

Table 17

Number of years with the firm by gender

Occupational Group	Mean no. of years		
	Women	Men	Total
Managers	—	9.1	8.7
Supervisors	6.4	10.0	8.4
White-collar (office)	9.9	4.8	6.3
Skilled	—	9.0	8.3
Factory	11.3	8.7	10.4
Total	8.0	8.7	8.5

[a]Dashes indicate less than ten cases: insufficient sample size to make an adequate comparison.

number of distinguishing features based on differing socioeconomic and class backgrounds. Multivariate techniques allow for consideration of a number of different influences on dependent variables. Stress is a complex area and involves a wide range of influences that need to be taken into account. Multivariate techniques are sophisticated enough to permit the treatment of the combined influence of a number of factors. Various writers have shown the benefits of multivariate techniques such as multiple regression for sociological research (e.g., 303–305).

A pooled analysis was used to test the effects of all independent variables together. However, effects of variables such as negative work, stress, low self-esteem, and poor mastery may vary for different occupations or for women and men; therefore, separate analyses have been conducted for major occupational groups and for men and women. As a test of labor process theory on stress and ill-health, the effects of control are derived through occupational membership (and to a lesser degree through gender). Separate analyses are an alternative to using interaction terms in pooled analyses.

A number of measures for variables and scales were developed in the study. A small number of the variable measures are nominal or ordinal. (Partnership is an example of a nominal variable in that no directional measurement scale is formed based on points in a scale. Occupational position is an example of an ordinal variable, with a scale from lowest to highest but without fixed intervals between points in the scale.) It is common practice in conducting survey research, particularly on a topic as complex as occupational stress, to have a mixture of measurement scales. Because multiple linear regression requires independent variables to be measured on a continuous scale (interval or ratio), the technique of "dummy" variables has been used. This allows for comparisons within the variable rather than a score for the variable's variance against the dependent variables. For instance the use of "dummy" variables for occupational position compares all other occupational groups against the regression coefficient for the category "supervisors," the omitted category. On the scale the category of supervisor falls between managers, those with most influence, and lower level white- and blue-collar workers. Another example is for educational level, where the "dummy" variables are measured against secondary education, the omitted category.

A path modeling technique of Path Analysis is applied for all the major dependent variables. I use the multivariate technique of multiple regression and analyze the relationships between data based on a theoretical model of causal relationships. The technique and its assumptions are described in more detail later in this chapter.

NEGATIVE WORK EXPERIENCE

Chapters 3 and 4 included reviews of research which showed that management control of the labor process can produce disadvantageous work experience, especially for blue-collar workers. According to Schwalbe and Staples (7), job routine and job control are the chief work characteristics that can explain changes in labor process organization over work. They found occupational differences in work experience between blue- and white-collar groups. Caplan and coworkers' (83) study of occupational differences in job stress and strain also reported difference between blue- and white-collar workers, but on a wide variety of work factors. For most job characteristics, administrators reported the least negative work experience, and blue-collar workers such as machine hands and assemblers the most negative experience.

In this chapter I present the experience of negative or alienating work characteristics for workers in the study sample. In Chapter 8 I examine the major effects of work factors on stress; in Chapter 9, their effects on the sociopsychological states of low self-esteem and low personal mastery; and in Chapters 10 and 11, the effects of negative work factors together with those of stress, low esteem, and poor mastery, or ill-health and psychiatric impairment.

The sample is from one firm and in no way does the study attempt to generalize to the wider population. However, it is relevant as a large sample of a large manufacturing organization. While significance levels (<.05) are presented, these are not intended as a level of significance to generalize to the wider population.

The negative work experience scale is an additive 32-item scale, scored from 0 to 100 points. Each item of negative work comprising the scale has a four-point response category: 0, meaning never experienced; 1, moderately often; 2, very often; and 3, always experienced. The scale has been summed and converted to a 0–100 point scale. A number of studies of occupational stress such as that of Caplan and associates (83) employ scales of a wide variety of work aspects.

The reported experience of negative (or alienating) work is moderate (mean score 40.3 on a scale of 0–100). The range of mean scores is from 35.1 for managers to 50.3 for factory workers. While all white-collar categories (managers, supervisors, and white-collar (office) workers) are below the sample mean, blue-collar groups (skilled and factory) are above. Managers and white-collar office workers have the lowest scores for negative work. The F score of 15.94 reveals large differences in the experience of alienating work by occupational groups. Differences are significant between factory workers and all other groups, and between skilled workers and two groups, managers and white-collar workers (Student-Neuman-Kuels test, SNK). The SNK has been used rather than more conservative tests (e.g., Scheffe) as it is the most power efficient post-hoc test and reduces the possibility of making a Type II error—that is, of not finding a difference between groups when a difference exists. Each discrete gender group has significant differences in alienating work scores. However, differences in the experience of negative work between women and men are negligible except between factory women and men, where women report a higher negative work score.

Table 18 shows gender and occupational differences for the negative work experience scale. The scale breakdown into categories of low, moderate, and high is based on the sample score's range rather than on the scale response range. The low scores represent the responses in the bottom third of the response range; the moderate scores, the middle third of the response range; and the high scores, the top third of the response range.

Table 19 shows percentiles for occupational and gender categories within occupational groups. White-collar workers and managers have scores for the 25th percentile in the lowest ranges of the scale, while factory workers score in the highest ranges. The median alienating work score (50th percentile) for factory workers is the midpoint of the scale (50.0). This is well above the median for managers and white-collar workers (both 36.5), meaning that they experience alienating working conditions a little. No real differences occur between 25th percentile scores for women and men in all occupational groups except for factory workers (women 43.8, men 34.9): the bottom quarter of women have a much higher level of alienating work. Similarly, the alienating work score for the top three-quarters of factory women (75th percentile) is much higher than for the same group of factory men.

Determinants of Negative Work

A regression model is used to identify the key elements of alienating work. Table 20 presents the total effects of a number of variables on a scale of negative work experience. The variables included in the regression model are ascribed and achieved status variables, entered first in the regression equation. These variables precede occupational position as well as length of time in the one position: this

Table 18

Negative work experience: mean scores by occupation and gender[a]

	Mean	(S.D.)	n
Managers	35.1	(8.5)	47
Supervisors	39.4	(8.8)	44
White-collar (office)	36.1	(10.7)	39
Skilled	43.5	(10.3)	24
Factory	50.3	(10.5)	37
Total	40.3	(9.7)	191

$F = 15.94$; significant $< .01$

	Women			Men			
	Mean	(S.D.)	n	Mean	(S.D.)	n	t
Managers	32.8	(6.0)	8	35.6	(9.0)	39	−.84
Supervisors	41.2	(8.8)	19	38.0	(8.6)	25	1.22
White-collar (office)	35.9	(9.3)	29	36.6	(14.6)	10	−.16
Skilled	45.8	(6.6)	4	43.1	(10.9)	20	.48
Factory	53.3	(10.4)	21	46.4	(9.6)	16	2.06*
Total			81			110	1.62

$F = 4.41$; significant $< .01$ $F = 13.45$; significant $< .01$

$*p < .05$
[a]Differences in numbers for the whole sample and the sum of each of the occupational and gender groups are due to missing cases. The *t* values are for gender differences.

group of variables is entered second. The standardized regression coefficients displayed are total effects. They are the sum of the direct effects of the variable in the regression *plus* the indirect effects.

The direct effect is when the variable is entered controlling for the effects of all other variables. The indirect effects are gained via effects through intervening variables on the dependent variable. The product of all of the standardized partial regression coefficients between the measured variable, intervening variables, and the dependent variable is the indirect effect. In the regression equation, occupational position and years in the one position are the intervening variables.

The regression presented in Table 20 shows the total effects of ascribed and achieved statuses, occupational position, and length of time in the one position on alienating work. Partnership status is measured by "dummy" variables. Having no partner and having a partner not working are measured against the omitted category, working partner's score on alienating work. They are scored as comparisons of the working partner and do not represent a score directly for negative

Table 19

Percentile table of negative work experience by occupation and gender

	25th Percentile	50th Percentile	75th Percentile	(n)
Managers	28.1	36.5	41.7	(47)
Supervisors	34.6	39.1	44.8	(44)
White-collar (office)	29.2	36.5	42.7	(39)
Skilled	37.8	40.6	49.7	(24)
Factory	42.2	50.0	57.8	(37)
Total	32.3	39.6	47.9	(191)

	Women				Men			
	25th	50th	75th	(n)	25th	50th	75th	(n)
Managers	28.1	32.8	38.8	(8)	27.1	36.5	42.7	(39)
Supervisors	35.4	41.7	46.9	(19)	33.1	36.5	43.8	(25)
White-collar (office)	30.2	38.5	42.7	(29)	27.1	29.7	47.9	(10)
Skilled	39.8	45.8	51.8	(4)	35.9	40.6	49.0	(20)
Factory	43.8	54.2	62.5	(21)	34.9	49.5	53.4	(16)
Total	35.4	41.7	47.9	(81)	31.3	38.5	47.1	(110)

work. Educational level is also scored as a series of "dummy" variables, with primary, apprenticeship, and tertiary education included, and secondary education the omitted category. Occupational positions are also "dummy" variables, with supervisors the omitted category. Managers, white-collar (office) workers, skilled blue-collar, and unskilled factory workers are measured against supervisors. The work scale is scored 0–100.

A multivariate ANOVA model was tested for all variables that had high to moderate correlations against the dependent variables. No variables were found to have interaction effects.

Variations in negative work experience is explained moderately well by the variables in the model. While status variables explain low variance (R-squared = .06), occupational level and length of time in current position add a further 18%. This means that a total variance of .24 is explained by variables in the model. Factors apart from those in the model explain about three-quarters of variations in experiencing alienating work. Some of these variables, such as education and occupation, indicate class location. They show an influence on negative work experience.

Table 20

Determinants of negative work experience: regression coefficients

Independent variables	Dependent variable, negative work	
	b	beta
Ascribed and achieved statuses		
Age	0.83	.08
Primary education	1.54	.02
Apprenticeship	−3.32	−.11
Tertiary education	−6.63	−.28*
Gender	−0.03	−.01
Having no partner	2.79	.11
Nonworking partner	−1.50	−.05
Number of children	0.62	.06
Adjusted *R*-squared		**.06**
Occupational status		
Managers	−2.13	−.08
White-collar (office) workers	−3.00	−.11
Skilled workers	4.96	.15
Factory workers	9.08	.34*
Years in position	1.28	.14
Adjusted *R*-squared (includes all variables)		**.24**

*Significant at < .05

Differences in effects on negative work between levels of education are moderate (Figure 2). Compared with those having a secondary level of education, primary-educated workers are likely to have a little more than one and one-half (1.54) points (out of 100) more of negative work experience. Having a tertiary education is likely to lead to more than six and one-half (−6.63) points less of alienating work than having a secondary education. Most tertiary-trained employees occupy managerial positions. Their higher level of education is also likely to secure them jobs with less alienating content. Apprenticed workers are likely to experience about three points (−3.32) less of negative work experience than secondary-educated employees. No other status variables achieved a strong influence of alienating work.

Of the second block of variables, occupational differences accounted for the largest effect. Differences in unstandardized regression coefficients are shown in Figure 3 for each occupational category compared with supervisors, the omitted category.

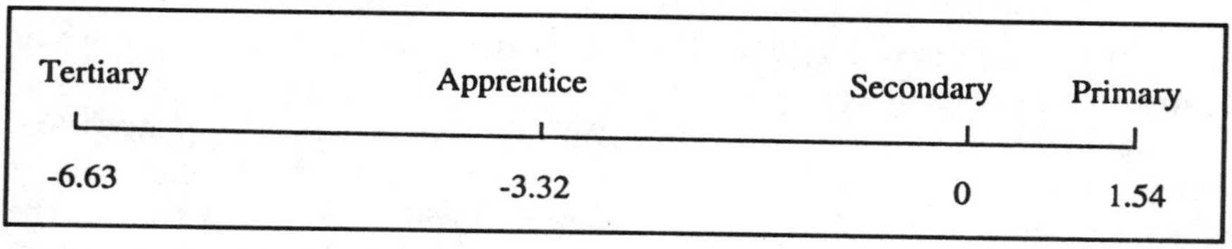

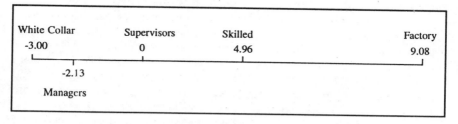

Figure 2. Unstandardized regression coefficients (*b*) scores: education and negative work.

Figure 3. Unstandardized regression coefficients (*b*) scores: occupation and negative work.

Differences between all white-collar groups (managers, supervisors, and white-collar (office) workers) and factory workers are large. Factory workers have the most negative work experience. A factory worker is likely to have about nine more points of alienating work than supervisors, and over 11 points more than managers. As the average alienating work score for the sample is 40.3, this represents a large disadvantage. Also, factory workers are likely to have about 12 more points of alienating work than white-collar workers. Skilled blue-collar workers have about five more points of alienating work than supervisors and almost eight more points than white-collar workers. White-collar workers are likely to experience more than two points less of alienating work than managers.

Workers with some tertiary education are most likely to have the least alienating work experience: those with middle-class educational advantage are likely to have better work experiences.

Negative Work Factors: Occupational and Gender Groups

Negative work factors are presented with mean scores for all work variables including all factor scales. Scores are presented for the factor scale. Mean scores for occupational groups are displayed, as well as scores for women and men. Standard deviations are given in parentheses. ANOVA scores are shown for occupational groups. Mean scores for scales and items are on a scale of 0–3. A score of 0 means never; 1, sometimes; 2, moderately often; and 3, always as scaled on the questionnaire.

Studies by Schwalbe and Staples (7), Caplan and associates (83), Karasek (65, 66), and others (reported in Chapters 2 to 4) show that blue-collar workers suffer some disadvantage in work experience compared with white-collar groups. The following results support this finding in nearly all cases: blue-collar workers perceive the most negative working conditions.

Component items of all the factor scales are presented in Chapter 6 in the factor analysis (Table 2, p. 113). A lack of consideration "represents an attitude toward employees by management based on its position of control over the work process, organization, and structure. Table 21 shows large occupational differences in "lack of consideration" ($F = 25.33$). While managers report some lack of consideration from management (.90), factory workers report the greatest lack of consideration (1.96). Differences exist between factory workers and all other groups, between skilled workers and all groups except factory workers, and between managers and supervisors (SNK test). In group interviews before the questionnaires were distributed, a number of factory workers said that management did not care for them and that managers maintained an attitude of superiority over lower level workers.

Table 21

Lack of consideration by management: mean scores by occupation and gender

	Mean (S.D.)	n			
Managers	.90 (.51)	53			
Supervisors	1.24 (.53)	45			
White-collar (office)	1.04 (.62)	41			
Skilled	1.62 (.66)	26			
Factory	1.96 (.59)	44			
Total	1.31 (.57)	209			

$F = 25.33; p < .01$

	Women		Men		
	Mean (S.D.)	n	Mean (S.D.)	n	t
Managers	.57 (.37)	9	.96 (.51)	44	−2.65*
Supervisors	1.38 (.51)	19	1.13 (.53)	26	1.54
White-collar (office)	1.11 (.62)	29	.88 (.62)	12	1.10
Skilled	1.92 (.96)	4	1.56 (.61)	22	.99
Factory	2.05 (.56)	28	1.80 (.62)	16	1.37
Total	1.45 (.75)	89	1.21 (.64)	120	2.35*

*p < .05

Table 22 shows that factory workers experience least work overload (1.07), which is well below the sample mean score. Skilled workers experience overload only occasionally. Managers report that they often experience overwork (2.70). The F score of 23.99 shows large occupational differences. Differences exist between managers and all other groups (white collar workers, factory workers, and skilled workers) and between supervisors, factory, and skilled workers (SNK test).

"Lack of opportunity to use skill discretion" includes the lack of opportunity to use previously learned skills and to acquire new ones on the job. It also includes the lack of opportunity to perform a variety of tasks. Karasek (65, 66) employs a similar measure, skill discretion, which includes both of these elements. This measure represents an aspect of management control over point-of-production activities. The mean score for Karasek's sample, which comprised a mixture of professional, administrative, white-collar, and blue-collar workers, was 2.68 (on a 1–5 scale), close to the scale midpoint. The mean score for this study is just below the scale midpoint (1.17 on a 0–3 scale) (Table 23).

As work moves from management to factory levels, the opportunity for skill discretion diminishes. As Table 23 shows, the one exception is that skilled workers report less lack of skill discretion (1.18) than white-collar workers (1.20). There are fewer opportunities for apprenticed work due to increased

Table 22

Work overload: mean scores by occupation and gender

	Mean (S.D.)	n				
Managers	2.07 (.52)	55				
Supervisors	1.63 (.58)	48				
White-collar (office)	1.70 (.74)	45				
Skilled	1.12 (.42)	27				
Factory	1.07 (.52)	46				
Total	1.57 (.57)	221				

$F = 23.99; p < .01$

	Women			Men		
	Mean (S.D.)	n		Mean (S.D.)	n	t
Managers	2.26 (.36)	9		2.03 (.54)	46	1.21
Supervisors	1.65 (.51)	20		1.62 (.63)	28	.18
White-collar (office)	1.55 (.66)	31		2.05 (.80)	14	−2.19*
Skilled	1.50 (.33)	4		1.06 (.40)	23	2.09*
Factory	1.09 (.56)	30		1.02 (.45)	16	.42
Total	1.49 (.65)	94		1.64 (.70)	127	−1.54

*$p < .05$

Table 23

Lack of skill discretion: mean scores by occupation and gender

	Mean (S.D.)	n
Managers	.72 (.41)	53
Supervisors	1.06 (.53)	46
White-collar (office)	1.20 (.61)	44
Skilled	1.18 (.59)	29
Factory	1.79 (.59)	46
Total	1.17 (.67)	218

$F = 24.78$; significant $< .01$

	Women			Men			
	Mean (S.D.)	n		Mean (S.D.)	n		t
Managers	.50 (.31)	8		.76 (.42)	45		−1.65
Supervisors	1.22 (.56)	20		.94 (.48)	26		1.82
White-collar (office)	1.23 (.63)	30		1.12 (.56)	14		.58
Skilled	1.08 (.57)	4		1.20 (.60)	25		−.36
Factory	1.88 (.60)	30		1.63 (.54)	16		1.40
Total	1.37 (.70)	92		1.03 (.57)	126		3.60*

$*p < .05$

mechanization, but the work that they perform has a strong skills base. Occupational differences are high ($F = 24.78$). Differences are between factory workers and all other groups, and between managers and all other groups (SNK test).

The quality of work relationships combines "not being treated with real feelings" and "tensions with people at work." It represents an aspect of management's control of structure. White-collar workers perceived the best work relationships (.91) (Table 24). Skilled workers, managers, and supervisors had a similar perception of work relations. Factory workers reported the poorest quality relationships (1.48). The F score for occupational differences is 5.55 and represents small to moderate differences. The major differences are between factory workers and all other groups (SNK test).

"Role conflict" represents an outcome of management control over the structure of work—that is, of bureaucratic control. It occurs in some measure because of limited discretion on the job: excessive direct management control can heighten conflicting role expectations. All occupational groups report role conflict to some degree. Table 25 shows that factory workers report experiencing role conflict moderately often (1.34). The F score is 2.40, representing only small differences between groups. At the factory level there is very little job discretion and workers must at times follow instructions from two or three different sources

Table 24

Poor work relationships: mean scores by occupation and gender

	Mean (S.D.)	n
Managers	1.07 (.57)	55
Supervisors	1.02 (.58)	49
White-collar (office)	.91 (.62)	43
Skilled	.95 (.63)	28
Factory	1.48 (.81)	46
Total	1.10 (.64)	221

$F = 5.55; p < .01$

	Women		Men		
	Mean (S.D.)	n	Mean (S.D.)	n	t
Managers	1.11 (.33)	9	1.07 (.61)	46	.22
Supervisors	1.07 (.46)	21	.98 (.66)	28	.53
White-collar (office)	1.00 (.67)	30	.69 (.43)	13	1.52
Skilled	1.25 (.29)	4	.90 (.66)	24	1.04
Factory	1.62 (.76)	30	1.22 (.84)	16	1.63
Total	1.23 (.67)	94	1.00 (.65)	127	2.69*

*$p < .05$

of authority, not just from their own foreperson or immediate supervisor. All groups report role conflict sometimes, however, showing that it is a common problem associated with management control throughout the organization. Differences are between factory and skilled workers only (SNK test).

Finally, "lack of freedom to control the job" is a single-item measure. Research by Frankenhaeuser and Gardell (109), Karasek (65, 66), and Schwalbe and Staples (7), among other studies reported in Chapter 2, shows that a perceived lack of control affects stress levels. It is a measure of control over the task. As shown in Table 26, factory workers report often lacking freedom to control work (1.78); skilled blue-collar workers also report a moderately high score (1.34). However, white-collar groups report lower scores, less than the sample mean (.92). The F score of 20.55 shows large occupational difference. Differences are between factory workers and other groups, and between skilled workers and all groups except factory workers.

In summary, for all negative work factors (the result of management's control over point-of-production activities, structure, and the climate of attitudes and values), factory workers experience the most negative, alienating experience. Occupational differences are large except for the effects on management control over structure.

Table 25

Role conflict: mean scores by occupation and gender

	Mean	(S.D.)	n
Managers	.96	(.64)	55
Supervisors	.94	(.63)	49
White-collar (office)	.92	(.85)	45
Skilled	.97	(.78)	29
Factory	1.34	(1.01)	47
Total	1.03	(.80)	225

$F = 2.40$; n.s.

	Women			Men			
	Mean	(S.D.)	n	Mean	(S.D.)	n	t
Managers	1.00	(.50)	9	.96	(.67)	46	.19
Supervisors	1.19	(.75)	21	.75	(.44)	28	2.58*
White-collar (office)	1.87	(.92)	31	1.00	(.68)	14	−.47
Skilled	.50	(.58)	4	1.04	(.79)	25	−1.30
Factory	1.42	(1.06)	31	1.19	(.91)	16	.75
Total	1.11	(.97)	96	.96	(.89)	129	1.39

*$p < .05$

Based on research reviewed in Chapter 2, women are expected to experience some differences in alienating work. However, in order to examine specific differences between gender groups, women and men are compared in the same occupational groups. Few gender differences are evident within occupational groups. Male managers report a greater "lack of consideration" than female managers. Among white-collar (office) workers, men report a greater level of overwork than women. While the greatest difference in "lack of skill discretion" between women and men is for supervisors, this difference is relatively small. Female supervisors also experience a moderately large disadvantage compared with their male counterparts in role conflict. Only minor differences occur between women and men in reporting poor quality relationships. Female supervisors and female factory workers report a lack of control over work moderately more than their male coworkers. Female supervisors experience some disadvantage with work factors relating to the task itself.

DISCUSSION

Many factors affect the experience of negative work and the likelihood of certain occupational groups being disadvantaged. Certain advantages are acquired

Table 26

No freedom to control work: mean scores by occupation and gender

	Mean (S.D.)	n
Managers	.38 (.53)	55
Supervisors	.75 (.84)	48
White-collar (office)	.61 (.87)	44
Skilled	1.34 (.90)	29
Factory	1.78 (1.13)	46
Total	.92 (1.10)	222

$F = 20.55; p < .01$

	Women			Men			
	Mean (S.D.)	n		Mean (S.D.)	n	t	
Managers	.33 (.50)	9		.39 (.54)	46	−.30	
Supervisors	1.05 (1.05)	20		.54 (.58)	28	2.18*	
White-collar (office)	.53 (.86)	30		.79 (.89)	14	.90	
Skilled	1.00 (1.15)	4		1.40 (.87)	25	−.82	
Factory	2.07 (.98)	30		1.25 (1.24)	16	2.45*	
Total	1.14 (1.14)	93		.77 (.86)	129	2.85*	

*$p < .05$

through increased education. Educational differences often reflect class differences. Those with higher (middle-class) educational advantages are likely to have jobs that impose fewer negative demands, and are more likely to be employed at higher levels and have more opportunities to be promoted to work with a less negative impact. They are likely to occupy positions that complement their skills training base. The effects of their having learned greater personal skills are likely to increase their ability to deal effectively with work pressures and with a lack of resources.

Even though labor process writers such as Braverman (2) have described the work of some white-collar jobs such as clerical and secretarial work as deskilled, these employees experience low levels of negative impact. One explanation is that they have not suffered the same degree of job degradation and lack of influence over work as blue-collar groups. White-collar work roles are usually regarded as supportive of management. Workers in these jobs generally support middle-class aspirations and have attitudes similar to those of managers. White-collar jobs often have a privileged status as they are normally the "feeding" ground for management. They consequently enjoy many of the working conditions and job characteristics of management, but with fewer responsibilities and smaller workloads.

In Chapter 3 I outlined Karasek's (66) argument for two traditions in stress research: one focusing on factors relating to the task itself, the other including other aspects of work experience. Alienating work factors in this study show that negative experience for blue-collar workers is in task-related job aspects, but is also high for work experience relating to management's control of structure and its effect on the climate of management attitudes and values. Braverman (2) argued that, as a result of Taylorism, management gained sufficient control over the planning and other discretionary elements of work. As management attitudes are based on exercising control that does not appreciate and value the work of its employees, this reflects a basic alienating characteristic of capitalist work experience.

Managers often measure their worth and effectiveness by the number of excessive demands placed on them. While their workload may be high, it can also be used as a measure of worth and of having a useful role to play. One component item of the scale, "frequency of taking work home," is most relevant to manager's work. Machine assemblers in Caplan and coworkers's (83) study reported little unwanted overtime at work. This may be related more to the need for extra income and therefore not minding extra work than to accepting extra workload demands per se. Much of the work of managers, according to Braverman (2), is planning and organizing, whereas much factory work is simply executing or doing the job. This difference affects the consequences of work overload for the different occupational groups.

The work of Braverman and other labor process writers on the consequences of Taylorism for deskilling has been outlined in Chapter 3. Increased management control has affected white-collar as well as blue-collar jobs, although job specialization has had the greatest effect on the lower occupational groups. Caplan and coworkers' (83) study found that assembly workers and forklift drivers suffered the greatest underutilization of skills. The less complex a job, the more were skills underutilized. A function of increased management control is the less complex work at lower occupational levels.

Schwalbe and Staples (7), using regression analysis, found that working-class (blue-collar) jobs were the most routine. In this study, factory workers reported learning new skills least often, using previously learned skills least frequently, and having least variety in their work. The results of deskilling and of work design processes such as Fordism are that manual and factory workers experience the least skill discretion in their jobs.

The quality of work relationships is affected by the type of workflow procedure, which in turn is influenced by the technology employed. People also bring to the job different expectations about the quality of work relationships. Former experience of poor work relationships can mean that workers lower their expectations about work relationships. Due to unfavorable working conditions, many blue-collar workers may expect fairly poor interactions. White-collar workers are more likely, through their past socialization and work experience, to expect satisfying relationships.

Role conflict can precipitate further limits in work discretion, making workers unsure of work requirements, and can result in management's control tightening. Caplan and associates (83) found a smaller spread of scores between occupational groups for this factor than for others. However, in their study there were clear differences in results between blue- and white-collar groups, and between manual and nonmanual workers.

High scores on lack of control for factory workers and low scores for white-collar workers were also shown in Caplan and coworkers' study. This is consistent with their findings for assembly workers (83). They found that among a diverse sample of occupational groups, the opportunity to participate in decisions was lowest for machine tenders and assemblers. Administrators had the highest opportunity, while white-collar workers were close to the mean for the sample in their study (2.88 on a 1–5 scale). Frankenhaeuser and Gardell (109) also reported low levels of control for assembly workers. Karasek records that in the United States, the "most common occupational codes with high levels of job demands and low levels of decision latitude [are] assembly workers" (65, p. 303).

The alienating effects of management control over task activities are reported more by women than men in the study, but the differences are moderate. Factory women are likely to satisfy fewer needs at work as the work role is most likely not their sole important role in life. Women tend to have a wider range of important out-of-work roles and therefore may be more prepared than many men to acknowledge the lack of need satisfaction in work tasks. Game and Pringle (302) argue that the authority structure of management control is likely to support men. Traditionally, factory jobs are men's jobs and are backed by an implicit masculine authority. This may affect women's avenues for expressing influence and control more than men's. Men, on the other hand, may be more likely to accept jobs with little influence. The large alienating content of work for blue-collar men may not be expressed by them as a problem. As noted in Chapter 2, the role of breadwinner may lead men to play down the fact that their jobs are inherently routine with little scope for exercising influence. In order to maintain self-esteem, they may in fact ignore the alienating effects of some of these lower level jobs.

CONCLUSION

The company selected for this study has had a fairly traditional history for a manufacturing organization. Technological development has led to some deskilling and job fragmentation, and the gender composition of the workforce is weighted toward women working in lower level occupational positions (this workforce characteristic is substantiated by Game and Pringle (302) and by

Baxter and coworkers (306)). The sample population reflects this disproportional gender distribution in the workforce.

As expected, the study found large occupational differences in the experience of negative work. Blue-collar workers, particularly factory workers, had the most negative experience associated with aspects of work over which management exerts a large degree of control. White-collar workers, especially managers and white-collar office workers, experienced some, but a smaller amount of alienating work. White-collar (office) workers experienced the greatest advantage. Management's control of work structure and control reflected in a climate of negative attitudes and values had outcomes shown to disadvantage blue-collar workers, especially factory workers. Blue-collar workers in the sample were greatly disadvantaged in both skill use on the job and scope and variety of tasks performed. Blue-collar workers also reported greater lack of opportunity to exercise freedom to control their work than management and other white-collar groups. Other labor process writers argue with Braverman (238, 244) that management has gained significant control over the labor process.

Studies by a number of stress researchers such as Caplan and coworkers (83), Schwalbe and Staples (7), and Kohn and coworkers (133) demonstrate substantial differences between occupational groups in the experience of alienating work. Negative work experience has been examined within the context of the labor process by a relatively small number of writers, including Navarro (6), Coburn (4), and others. These studies confirm the disadvantage to blue-collar workers of management's substantial control over the labor process. Demonstrated disadvantages in the opportunity for skill discretion and freedom to exert control over the job support Braverman's argument that deskilling and job fragmentation most strongly affect lower level workers. However, Braverman (2) viewed control only at the point of production. Littler and Salaman (242) argue that increased control over the labor process is expressed also through structural, bureaucratic control and broader aspects of organizational control. In addition to control over the task, management's control of structure affects work process and management practice. The results of this study show that the outcomes of management control extend well beyond effects on the task itself.

CHAPTER 8

Occupational Stress:
The Effects of Alienating Work

A number of variables have been tested for their impact on the causal model of stress. Occupational position is a major factor that creates differences in stress, and as a result the major analyses of the causes and experience of occupational stress will deal with the effects of occupational position. In this chapter I test the first study hypothesis that "as occupational status moves to the bottom of the hierarchy, occupational stress is expected to increase." Other aims of the chapter are to test specific negative task-related factors, based on the work of Schwalbe and Staples (7). These factors relate to point-of-production activities, which are central to Braverman's analysis of management control. Finally, I present a detailed discussion of the component factors in the causal model of stress.

A number of writers have shown that many factors, including negative work experience, contribute to occupational stress (see 7, 83, 101, 307). A small number of writers in the labor process tradition have shown that blue-collar workers experience the greater disadvantage in alienating work due to their position of limited control and influence over the labor process (4, 7, 308). Other writers such as Kohn and coworkers (133), Otto (134), and Anderson and coworkers (135) have shown similar results, although relatively few writers have studied the effects of stress on blue-collar workers. While Braverman (2) argued that point-of-production activities best emphasize management control, Tanner and coworkers (247) suggest that his focus on these areas as the basis of forming worker consciousness is inappropriate. Other labor process writers such as Littler and Salaman (242) and Thompson (238) argue that management control of structure and of wider organizational activities accounts for such alienating work experience. Management's increased control at these levels also leads to many unrewarding and alienating experiences for workers.

Stress results from people not being able to satisfy important needs at work. Many of these cases of stress result from workers having insufficient influence to effect changes in job organization, work process, and structure. They are unable to change work conditions so as to satisfy important needs, due to the extent of management's control over the work process.

The occupational stress scale is an additive 32-item scale scored from 0-100. Table 27 shows a relatively small range of stress scores, from a low of 23.36 to a

Table 27

Occupational stress: mean scores by occupation and gender[a]

	Mean (S.D.)	n
Managers	23.36 (14.50)	46
Supervisors	27.76 (14.03)	43
White-collar (office)	23.52 (15.87)	35
Skilled	23.71 (17.72)	25
Factory	35.07 (17.04)	36
Total	26.77 (15.60)	185

$F = 3.72; p < .01$

	Women			Men			
	Mean (S.D.)	n		Mean (S.D.)	n	t	
Managers	20.18 (8.35)	8		24.15 (15.48)	38	−.70	
Supervisors	27.36 (14.60)	19		28.08 (13.87)	24	−.17	
White-collar (office)	24.04 (15.96)	27		21.88 (16.51)	8	.33	
Skilled	29.17 (14.32)	3		22.96 (18.06)	22	.57	
Factory	38.65 (16.75)	20		30.60 (16.85)	16	1.43	
Total	28.45 (16.17)	77		25.57 (15.94)	108	1.22	
	$F = 3.43; p < .05$			$F = .87$; n.s.			

[a]Differences in numbers for the whole sample and the sum of occupational and gender groups for the stress (32-item) scale are due to missing cases. The t values are for gender differences.

high of 35.07. The narrowness of the range reflects the scale categories used rather than sample characteristics. (Few workers would be expected to report being stressed "all of the time" on more than a small number of items. Only in extreme cases would a worker average a "fair bit" of stress on all 32 negative items.) The sample mean is 26.77 (less than "sometimes experiencing stress"). Occupational differences are moderate ($F = 3.72$). Major differences are between factory workers and all other groups (SNK test). Managers, white-collar workers, and skilled blue-collar workers record the least stress on average. For managers and white-collar workers, their low stress scores are consistent with their low scores for alienating work. Blue-collar factory (unskilled) workers have the highest average stress score. This corresponds with their high score for alienating working experience.

Gender differences in stress are minimal for those in the same occupational groups. Differences between women and men are most pronounced among unskilled factory workers, with women having a high stress score. However, differences are not significant. Differences between women and men in the other occupational groups are minimal. The major differences for men are between factory workers and managers and white-collar workers (SNK test). For women there are no differences between groups.

In Table 28, percentile scores are presented for occupational and gender groups. Some large differences exist between the lowest quartile groups. The bottom quarter (25th percentile) of unskilled factory workers have a score of 21.1 or less. Skilled workers have the lowest score for the bottom quartile (10.9), almost half that of the unskilled group. In addition, there are some large gender differences between the bottom quartiles (low-scoring groups) of white-collar (office) and factory workers. Women have considerably higher scores than men in these groups.

STRESSFULNESS OF
INDIVIDUAL NEGATIVE WORK FACTORS

A comparison of alienating work scores and scores for stressfulness for each of the factors showed occupational responses to the stressfulness of work factors. The extent of correlation between work factors and stress was measured. A test of occupational and gender difference examined the effects of structural location on negative work experience and stress. Each negative work factor was correlated with a scale rating of stressfulness for that item using Pearson's product movement correlations. A high score on the stress measure indicates high

Table 28

Percentile table of occupational stress by occupation and gender

	25th Percentile	50th Percentile	75th Percentile	(n)
Managers	12.0	20.8	33.3	(46)
Supervisors	17.7	25.0	39.6	(43)
White-collar (office)	12.5	19.8	35.4	(35)
Skilled	10.9	18.8	34.9	(25)
Factory	21.1	34.9	46.9	(36)
Total	17.2	25.0	39.6	(185)

	Women				Men			
	25th	50th	75th	(n)	25th	50th	75th	(n)
Managers	11.2	21.9	27.9	(8)	12.0	20.8	34.6	(38)
Supervisors	13.5	25.0	39.6	(19)	19.3	25.5	38.8	(24)
White-collar (office)	13.5	19.8	35.4	(27)	7.0	18.2	32.8	(8)
Skilled	13.5	32.3		(3)	9.1	18.8	34.6	(22)
Factory	24.2	38.5	48.4	(20)	14.3	25.5	45.8	(16)
Total	17.2	25.0	39.6	(77)	12.5	23.4	36.2	(108)

stressfulness. The negative work scale is from 0 to 3, with a high score indicating the most negative work experience.

Employees were asked to rate the degree to which they experienced alienating work and the stress they felt from it. Table 29 shows the mean scores for each negative work factor and its stressfulness for occupational groups. Pearson's product movement correlations are given for each pair of mean scores.

For the total sample, correlations between all reported negative work factors and stressfulness achieve a probability level of < .05. The highest correlations are for a "lack of consideration," "poor relationships," and "role conflict." There are high correlations between the degree of "lack of consideration by management" experienced and the level of stress for that factor, and high correlations for all occupational groups for "poor work relationships" and "role conflict."

Table 29

Negative work experience and stressfulness of that work experience

Work factor	Scale scores and Pearson's product movement correlations[a]					
	Mgr.	Spr.	W.-C.	Skd.	Fty.	All
No consideration by management	.91	1.24	1.05	1.62	1.96	1.32
Stressfulness	.73	.97	.69	.70	1.30	.88
Correlation	**.65***	**.50***	**.53***	**.55***	**.49***	**.56***
Poor relationships	1.07	1.05	.93	.95	1.47	1.09
Stressfulness	.94	.96	.82	.79	1.40	.98
Correlation	**.60***	**.81***	**.81***	**.63***	**.74***	**.73***
Lack of skill discretion	.72	1.04	1.21	1.18	1.79	1.17
Stressfulness	.61	.79	.74	.75	.91	.75
Correlation	**.29***	**.20**	**.48***	**−.32**	**.23**	**.23***
Overwork	2.07	1.63	1.71	1.12	1.07	1.59
Stressfulness	1.04	.85	.83	.57	.79	.86
Correlation	**.20**	**.43**	**.66***	**.66***	**.70***	**.53***
Role conflict	.96	.94	.95	.97	1.34	1.03
Stressfulness	1.02	1.17	1.05	1.03	1.36	1.14
Correlation	**.70***	**.71***	**.73***	**.72***	**.77***	**.73***
No freedom to control	.35	.77	.62	1.34	1.24	.92
Stressfulness	.39	.57	.60	.83	1.80	.70
Correlation	**.15**	**.30***	**.56***	**.39***	**.16**	**.39***

[a]Mgr. = managers, Spr. = supervisors, W.-C. = white-collar (office), Skd. = skilled, Fty. = factory.
*p < .05

Both "lack of skill discretion" and "lack of freedom to control work" are measures of alienation at the task level of work. For each of these work factors, several occupational groups recorded low correlations between level of alienation and stressfulness experienced. White-collar workers reported a mean score for skill discretion of 1.21, with a moderate to high correlation with the level of stress experienced. Managers reported a moderate correlation but supervisors, skilled workers, and factory workers a low correlation between skill discretion and stress experienced. For managers, "lack of skill discretion" opportunity was low: this was a less important source of stress than other negative work characteristics. Factory workers, supervisors, and skilled workers reported to moderate levels of underutilized skills and lack of task variety. For factory workers and supervisors this produced a relatively small amount of stressfulness: surprisingly, for skilled workers the negative correlation means that a lack of skill discretion is a source of low stress, but this is not significant.

For supervisors, white-collar workers, and skilled workers there is a moderate to high correlation between a lack of freedom to control work and stress experienced. Skilled workers experienced a relatively small degree of freedom to exercise control (mean 1.34); they reported being stressed a little by this (.83). There are low and insignificant correlations for managers and factory workers. Managers have virtually total freedom to control (.35) and reported a small level of stress. Factory workers reported being stressed often (1.80) by the lack of opportunity for control.

There are only small and insignificant associations between level of "workload" and stress for managers and supervisors. Other groups show high correlations. Managers perceived overwork to a high degree (2.07) yet were stressed only occasionally (1.04).

As Table 30A shows, for female managers there are no significant correlations between negative work factors and stressfulness: alienating work characteristics are not associated with a lack of need satisfaction. For female supervisors, however, all negative work factors, those relating to the task, effects of management control on structure, and effects on the climate of negative attitudes by management, are accompanied by fairly closely related stress levels. High alienation is likely to produce high stress. There is a very low and insignificant correlation between overwork and its stressfulness. Only a lack of consideration by management has no correation with stressfulness for white-collar women. All other work factors produce a fairly highly related stressfulness. The number of skilled women is too low to draw any firm conclusions from the sample. Factory women, however, reported a moderate to high stress and alienating work correlation for all aspects of work. They have the highest levels of alienating work, and this satisfies few important needs for them.

Table 30B shows that for male managers, there is a moderate to high correlation between stressfulness and the effects of management control on structure and a negative climate of attitudes by management. However, stressfulness is not related to the experience of negative task factors and overwork. Lack of freedom to control and skill discretion is not likely to lead to stress. Male supervisors reported moderate to strong correlations only between stressfulness and the effects of structure (role conflict and pool relationships) and overwork. Similarly, there is a strong correlation of stress with only one element of the effects of control over structure: poor relationships. Skilled blue-collar men have no significant correlation between task-related activities and their stressfulness: for these employees, all other work factors produce a high level of stress for high alienation. Finally, a lack of skill discretion, a lack of control, and overwork have no significant correlation with stress for factory workers. If these alienating characteristics are evident in their work, they do not represent important sources of need satisfaction.

Table 30A

Negative work experience and stressfulness of that work experience: women

Work factor	Scale scores and Pearson's product movement correlations					
	Mgr.	Spr.	W.-C.	Skd.	Fty.	All
No consideration by management	.57	1.38	1.11	1.92	2.05	1.44
Stressfulness	.43	.95	.69	.79	1.39	.93
Correlation	.65	.70*	.62*	−.01*	.36	.60*
Poor relationships	1.11	1.10	1.00	1.25	1.60	1.23
Stressfulness	1.22	1.13	.87	1.63	1.53	1.19
Correlation	.58	.80*	.86*	.98*	.69*	.77*
Lack of skill discretion	.50	1.22	1.23	1.08	1.88	1.36
Stressfulness	.21	.77	.76	.92	.94	.77
Correlation	−.52	.62*	.55*	.66	.14	.45*
Overwork	2.26	1.65	1.58	1.50	1.09	1.51
Stressfulness	1.30	.88	.79	.92	.92	.91
Correlation	.62	.16	.75*	.68*	.83*	.61*
Role conflict	1.00	1.20	.90	.50	1.42	1.11
Stressfulness	1.33	1.35	.93	1.00	1.48	1.23
Correlation	.58	.67*	.84*	.82	.85*	.79*
No freedom to control	.33	1.05	.55	1.00	2.10	.79
Stressfulness	.33	.60	.55	.50	1.36	1.16
Correlation	.50	.66*	.60*	1.00	.11	.51*

*p < .05

In Chapter 3 I described a specific problem with alienation: even though workers may experience negative work, many may not feel a lack of important need satisfaction. For men in the sample there was no significant correlation between task activities and associated stressfulness. Most groups of men, however, showed a correlation between alienating effects of control over structure, the negative climate of attitudes by management, and associated stressfulness.

For female managers no work factors relate to important need satisfaction. In contrast, factory women reported high correlations between stress levels and all negative work experiences. Stressfulness associated with task activities is characteristic of most groups of women, as is stressfulness associated with the effects of management control on structure and the climate of negative attitudes.

Table 30B

Negative work experience and stressfulness of that work experience: men

Work factor	Scale scores and Pearson's product movement correlations					
	Mgr.	Spr.	W.-C.	Skd.	Fty.	All
No consideration by management	.98	1.14	.88	1.56	2.80	1.23
Stressfulness	.80	.98	.69	.68	1.46	.85
Correlation	**.62***	**.39**	**.37**	**.73***	**.66***	**.52**
Poor relationships	.88	1.02	.73	.90	1.22	1.00
Stressfulness	1.07	.83	.68	.65	1.56	.84
Correlation	**.61***	**.85***	**.61***	**.61***	**.81***	**.69***
Lack of skill discretion	.76	.91	1.17	1.20	1.63	1.05
Stressfulness	.68	.81	.69	.72	.83	.73
Correlation	**−.28**	**−.08**	**.19**	**−.38**	**.40**	**.04**
Overwork	2.03	1.62	2.03	1.06	1.02	1.63
Stressfulness	.99	.83	.94	.51	.54	.81
Correlation	**.13**	**.61***	**.53**	**.61***	**.32**	**.48***
Role conflict	.96	.74	1.08	1.04	1.19	.96
Stressfulness	.96	1.04	1.31	1.04	1.13	1.05
Correlation	**.73***	**.76***	**.43**	**.75***	**.55***	**.65***
No freedom to control	.36	.56	.77	1.40	1.25	.64
Stressfulness	.40	.56	.69	.88	1.00	.77
Correlation	**.10**	**−.18**	**.49**	**.32**	**.13**	**.27***

*$p < .05$

DETERMINANTS OF OCCUPATIONAL STRESS

Based on the causal model presented in Chapter 6, occupational stress has a number of determinants. In Chapters 2 and 3 I reviewed research which showed that certain alienating or negative work experiences are especially conducive to occupational stress. The purpose of the regression analysis shown in Table 31 is two-fold: first, to ascertain the major status, occupational, and alienating work determinants of stress; second, to present the first major hypothesis, that those in lower occupational positions are expected to experience greater stress.

Table 31

Determinants of occupational stress: regression coefficients (total causal effects)

	Dependent variable, stress	
Independent variables	b	beta
Ascribed and achieved statuses		
Age	−0.55	−.04
Primary education	2.79	.03
Apprenticeship	−4.36	−.09
Tertiary education	−6.33	−.18*
Gender	−0.86	−.03
Having no partner	4.46	.13
Nonworking partner	0.28	.01
Number of children	1.97	.13
Adjusted R-squared		**.02**
Occupational status		
Managers	−1.29	−.03
White-collar (office) workers	−4.13	−.10
Skilled workers	−3.37	−.07
Factory workers	5.89	.15
Years in position	0.85	.06
Adjusted R-squared		**.04**
Alienating work		
Role conflict	5.58	.28*
Poor relationships	5.75	.24*
No freedom to control work	−0.75	−.05
Overwork	3.10	.13
Lack of skill discretion	3.05	.12
Lack of consideration by management	8.71	.37*
Adjusted R-squared		**.41**

*Significant at < .05

Variations in occupational stress scores are explained moderately well by variables in the model. With all variables entered, 41% of variability (R-squared = .41) in stress scores is explained. However, the first block of variables, ascribed and achieved statuses, occupational status, and length of time in the current position, explain only 4% of variability. With the addition of negative work factors (the second block of variables), a large amount of variance is explained by the model. Table 31 shows the total effects of ascribed and achieved status, occupational position, job experience, and work characteristics. Interaction effects for gender, level of education, and occupation were tested through multifactorial ANOVA. No interaction effects were found to be significant for stress.

Differences between level of education are generally small. Primary-level education, apprenticeship, and tertiary education are compared against the omitted category of secondary education. Figure 4 shows that the major difference is between those with primary education only and those with some tertiary education (the tertiary-educated have more than 9 points less stress (out of 100) than primary only). In the sample, most employees with tertiary education worked in management positions. Their higher educational level has given them greater opportunities to work in jobs with less alienating work characteristics. However, small and insignificant differences in stress exist only with differences between partnership status and number of children.

In Chapter 6 I presented four major hypotheses for testing (p. 105): three of these test the effects of occupational position on four of the dependent variables. The first hypothesis tests for structural effects on stress. Based on a review of literature in Chapters 2 to 5, we would expect differences in stress levels at work between all white-collar groups and blue-collar workers, especially factory workers: "As occupational status moves toward the bottom of the hierarchy, occupational stress is expected to increase." Studies by Caplan and associates (83), Schwalbe and Staples (7), and others have shown that stress levels for work conditions experienced by managerial groups and other white-collar groups are lower than those experienced by blue-collar and especially unskilled workers. The review of labor process stress writers (4, 6, 8) in Chapters 3 to 5 showed that

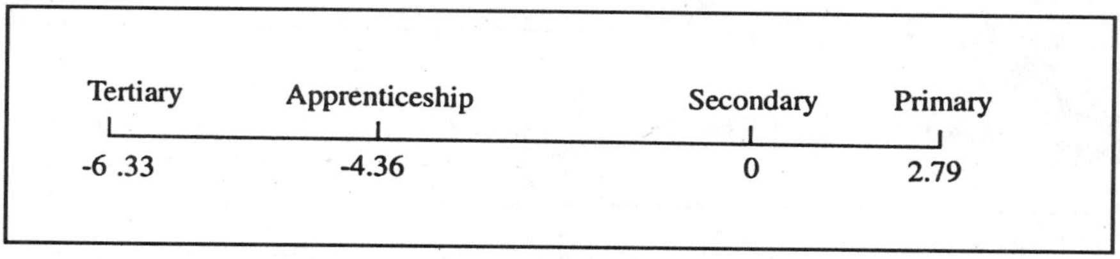

Tertiary	Apprenticeship		Secondary	Primary
-6 .33	-4.36		0	2.79

Figure 4. Unstandardized regression coefficient (b) scores: educational level and stress.

workers in lower occupational positions are expected to have fewer opportunities for exercising control over their jobs and influencing work processes.

The second block of variables entered in the model included occupational position. Occupational position is comprised of dummy variables, with each occupational group compared with supervisors, the omitted category. Figure 5 shows unstandardized regression coefficient scores comparing stress levels for each occupational group.

The hypothesis that as occupational position decreases, stress levels increase, cannot be completely supported. White-collar workers and skilled workers can expect to experience least stress. Managers are likely to experience more stress than both of these groups; factory workers are likely to experience the greatest amount of stress. There is an increase of more than 10 points of stress (out of 100) from white-collar (office) positions to factory jobs, and almost as great an increase in stress points (more than 9 points) between skilled workers and factory workers.

The third block of variables entered in the model includes separate negative working conditions. The major negative work determinant of stress in the sample is a "lack of consideration by management." The standardized regression coefficient (beta) of .37 shows a large influence on stress levels for those who experience lack of appreciation and consideration by management. "Role conflict" (beta = .28) is also a moderate determinant of stress, and "poor relationships" has a moderate effect (.24).

OCCUPATIONAL GROUPS AND STRESS

In order to test the effects of structural location on stress, separate multiple regression equations were calculated for white-collar workers and blue-collar workers. Again, three separate blocks of variables were entered. The first included ascribed and achieved statuses, together with length of time in the one

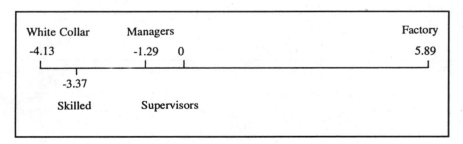

Figure 5. Unstandardized regression coefficient (*b*) scores: occupational position and stress.

job. These variables did not produce sufficiently large effects and
shown. Table 32 shows the effects of all negative work factors.

Certain work factors are more conducive to stress for blue-collar (working-class) occupational groups than for white-collar (middle-class) workers. The *R*-squared for blue-collar workers is .54; the negative work factors alone explain 42% of variability in their stress scores. For white-collar workers the *R*-squared is .32.

For white-collar workers, "poor work relationships," an outcome of management control over the structure of work processes, is the major determinant of stress (beta = .33). "Lack of consideration by management," an element of the organizational climate reflecting management's control, is the second major determinant of stress (.22). "Lack of skill discretion" (.21) is also significant. White-collar workers (managers, supervisors, and white-collar office workers) are likely to expect moderately satisfying work relationships. Among blue-collar workers, women report more stress than men (beta = −.33). "Poor work relationships" is not an important determinant of stress for blue-collar workers, for whom "lack of consideration from management" has a major effect (.39). "Role conflict" is the other alienating characteristic with a large effect on stress (.33) for blue-collar workers.

GENDER AND STRESS

The major determinants of men's and women's experience of stress were tested in two separate regressions. The aim was to look for variations between women

Table 32

Determinants of occupational stress by occupational group

	Dependent variable, stress (beta)	
Independent variables	White-collar	Blue-collar
Gender	.06	−.33
Negative work factors		
Role conflict	.16	.33*
Poor relationships	.33*	.14
No freedom to control	−.00	−.06
No skill discretion	.21*	.10
Overwork	.09	.15
Lack of consideration	.22*	.39*
Adjusted *R*-squared	**.32**	**.54***

*Significant at < .05

and men in conditions conducive to stress. Table 33 shows that for women, a moderate to large amount of variance in stress scores is explained by factors in the model (R-squared = .48); for men, small to moderate variance (.33) is explained.

For men, having a tertiary education is likely to lead to lower stress scores. For women, working in blue-collar compared with white-collar positions is likely to lead to large differences in stress scores. For both women and men, role conflict (conflicting expectations about work requirements) is a moderate determinant of occupational stress. "Poor relationships" at work and a "lack of skill discretion" are the other negative work determinants for women. For men, the other major determinant of stress is "lack of consideration by management."

A PERCEIVED LACK OF CONTROL AND JOB ROUTINE

The effects of "a perceived lack of control" and "job routine" on occupational stress were included to review their role as found by Schwalbe and Staples (7) and by various other authors. Schwalbe and Staples used "lack of control" and "job routine" as important outcomes of the organization of the labor process for the nature of work for lower level employees. Elements of management control other than task variables deal with control of structure and climate, and were found earlier in this chapter to have a stronger effect than task-related negative factors.

Table 33

Determinants of occupational stress by gender

Independent variables	Dependent variable, stress (beta)	
	Women	Men
Tertiary education	−.14	−.21*
Occupational position	.35*	−.04
Negative work factors		
Role conflict	.26*	.26*
Poor relationships	.36*	.14
No freedom to control	−.01	−.07
No skill discretion	.28*	−.01
Overwork	.12	.12
Lack of consideration	.25	.40*
Adjusted *R*-squared	**.48**	**.33**

*Significant at < .05

In this study, two factors, "lack of control" and "job routine," were developed in a factor analysis and were reported in Chapter 6 (Table 4, p. 118). A least squares regression equation was calculated. The joint effects of a lack of control and job routinization are entered as the third block of variables. Ascribed and achieved statuses, occupation level, and job experience are the same as for the regression shown in Table 31 and hence are not reported. As Table 34 shows, "lack of control" has a greater impact (beta = .26 and significant) than "job routine" (.21, not significant). Twelve percent of variability in stress scores is explained in the model; these variables therefore explain little of the variability in stress scores. Schwalbe and Staples (7) found a similar result for these two task-related variables. A "lack of control" achieved a beta score of –.26 (significant), while "job routine" remained lower at –.16 (significant). Thus, through the use of different measures of the variables, the current study confirms their findings on the effects of these two task-related negative work factors, considered without other relevant work factors.

Lack of Control and Job Routine:
Occupational and Gender Differences

The influences of occupational position and gender were tested for their effects on these two task-related variables. Results from the regression equations for white-collar and blue-collar employees and for women and men are presented in Table 35. For white-collar groups (managers, supervisors, and white-collar office workers) only job routine has an effect: it is moderate. The reverse is the case for blue-collar workers: job routine has a minor effect on stress compared with a perceived lack of control. For men, a lack of control and job routine have no influence. However, a perceived lack of control is important for women.

Table 34

Effects of lack of control and job routine
on occupational stress

Independent variables	Dependent variable, stress (beta)
Lack of control	.26*
Job routine	.21
Adjusted *R*-squared	**.12**

*Significant at < .05

Table 35

Effects of lack of control and job routine on stress:
occupational and gender differences

Independent variables	Dependent variable, stress (beta)			
	White-collar	Blue-collar	Women	Men
Lack of control	.19	.32*	.40*	.22
Job routine	.27*	.11	.25	.06
Adjusted *R*-squared	**.08**	**.21**	**.22**	**.02**

*Significant at < .05

PATH MODEL OF OCCUPATIONAL STRESS

In Chapter 6 I described the causal model for stress and ill-health. The negative work dimensions discussed in the path model are the three dimensions that resulted from the factor analysis of factors presented in that chapter (Table 3, p. 116) rather than the six negative work factors. A path diagram (Figure 6) describes the components of the causal model leading to stress. It graphically represents the strength and direction of relationships between variables in the model of stress and the various causal relationships between variables. Four groups of variables were entered in sequence, with the first independent variables on the left of the diagram, progressing through to the dependent variable, stress, on the right. Standardized regression coefficients (beta) are given as the causal weights for each path. For second- and third-block variables and the dependent variable, residual paths are shown by a vertical arrow. These represent effects not explained in the model on the particular variable.

The model is composed of ascribed and achieved statuses, occupational status, and job experience. It also includes three dimensions of negative work experience. These all combine to explain the various influences on the dependent variable, stress. The path model shows the strength of the varying causal relationships. Only those causal paths with regression coefficients at < .05 significance are displayed. In reality, paths exist between all variables.

Five ascribed and achieved status variables have been entered in the model, but only two are shown. Causal paths from gender and number of children to other variables in the model are not large enough to be included. The path from age does not continue beyond number of years in the one job.

Education has a direct effect on stress (beta = −.14). One indirect path to stress is given in the diagram, via occupational position and via management control over task and negative climate (−.49 × .55 × .20 = −.05). The total effect of educational attainment is the sum of direct and indirect effects. The total effect

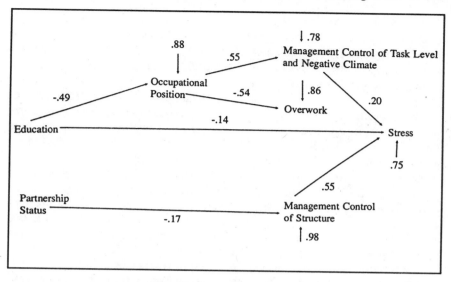

Figure 6. Path model for stress (including dimensions of negative work experience).

(−.14 + (−.05) = −.19) indicates that a lower education is likely to lead to greater stress. The total effect, however, does not include the indirect effects of other variables whose paths have not been included in the path model. The main purpose of the path model is to show both the causal paths and directions and how the total effects are calculated.

Partnership has an effect only on the outcome of management control of structure (beta = −.17). This effect is small and indicates that having a working partner compared with no partner is likely to lead to a more negative experience of poor work relationships and role conflict, two outcomes of management control of structure. Partnership has a very small indirect effect on stress via management control of structure (−.17 × .55 = −.09). The direct effect of partnership on stress is not sufficiently large to display in the path diagram. The total effect of partnership on stress is −.11.

Occupational position is an important variable in this model. A small amount of what determines occupational position is explained by preceding variables in the model, as shown by the large residual .88). Occupational position has two effects recorded in the diagram. The first path is to management control of task and negative management attitudes (.55). This means that blue-collar work leads to experiencing the most negative effects of control over task and negative management attitudes. The second path is to overwork (−.54): being a manager is most likely to lead to experiencing overwork. Occupational level has one indirect path to stress via management control of task level and negative climate of management attitudes (.55 × .20 = .11). The path coefficient of .20 shows a relatively small effect of management control of task and of negative attitudes on experiencing stress. The total effect on stress of occupational position is the sum of direct and indirect effects (−.03 + .11 = .08). The direct effect is not large enough to be shown in the diagram.

There is little in the model explaining management control of structure: the residual is .98. The causal path to stress (beta = .55) is large, showing that outcomes of management control on structure (poor relationships and role conflict) have a very important effect on stress. Poor relationships and conflicting role expectations can be expected to lead to high stress. Negative effects of management control of task and poor management attitudes, however, have a relatively small effect on stress (.20) and are less important.

Overall the model predicts that high stress is most likely to be an outcome of low levels of education and of blue-collar rather than white-collar work. It is also likely to be an outcome of work that has role conflicts and poor relationships. To a lesser degree stress will also result from work that offers little control and skill discretion and where management expresses negative attitudes toward workers. Each of the latter elements is most characteristic of working-class, blue-collar jobs.

The causal model confirms previous research reported and discussed in Chapters 3 and 4. This includes research by Schwalbe and Staples (7), Caplan and associates (83), and others who show that negative work characteristics belong more to blue-collar than white-collar occupations and are a major cause of occupational stress. The findings of this study show also that management control over the task and workers' lack of opportunity to use skills and have varied tasks, together with negative management attitudes, are overshadowed by other measures of management control, particularly outcomes of the control of structure. Factors related to management's control of the work environment rather than the job itself are the most evident in causing occupational stress.

DISCUSSION

In Chapters 3 and 4 I argued that occupational position affects the degree of control over the labor process. Also, in Chapter 7 I showed that alienating work factors can measure the degree to which management exerts control over the work environment. Gender within occupational group is less well researched on the effects of stress differences.

Not all studies of stress use a measure of stress. Those that do have large variations in the way in which the stress concept is measured. Schwalbe and Staples (7) employ a scale of negative emotional reaction to work experience. They report a mean score of 18.33 (S.D. = 4.59) on a possible score range of 10–40. This represents an average of almost a third of the possible low score for the sample in this study. Caplan and coworkers (83) also use a measure of negative emotional reaction, and they report higher stress levels than Schwalbe and Staples. Other writers, such as Anderson and coworkers (135), have studied the effects of stress on different occupational groups. They found that front-line health workers had higher stress than either doctors or nurses.

The low scores for stress for managers and white-collar workers in this study relate to low scores for alienating working conditions. Managers usually enjoy a moderate to large degree of discretion and control over their work environment. They experience high appreciation for the work they do from management, and usually experience little alienation and lack of need fulfillment. There are not many studies comparing composite groups of employees and managers, although some, such as that of Thorell and colleagues (309), look at selective occupational categories. Other studies on occupational stress such as Caplan and coworkers' (83) study compare composite groups such as managers, administrators, and professionals. Very few studies compare managers and other white-collar groups with blue-collar groups.

The white-collar workers (such as secretaries, sales staff, computer operators, and accountants) in our sample reported well below the sample mean for negative work experience. These are generally middle-class persons and as such probably expect to satisfy important needs at work, due to previous positive work experiences and good levels of education. These positions are fairly high in autonomy and skill utilization (see Chapter 7). Promotion opportunities in the company were good, and many of these white-collar employees could expect to move into management.

Blue-collar skilled and semi-skilled workers reported one of the lowest average stress scores while recording one of the highest scores on alienating work. Their mean negative work experience score was well above the average (see Chapter 7). In Chapter 3 I outlined Braverman's (2) argument that skill levels have declined dramatically for blue-collar workers as a result of management's increased control over the labor process. Skilled blue-collar work, however, gives workers a basis for a sense of competence, despite their reports of lack of appreciation by management. This could explain how the satisfaction of important needs overrides some of the effects of negative work. Even though many aspects of their work provide negative work experience, especially a lack of appreciation by management and a lack of freedom to control work, skilled trade workers have managed to preserve a basis of craft identification and craft membership.

For almost all negative work experience factors, factory workers have a high score. These factors include a lack of appreciation, little opportunity for skill discretion, poor work relationships, limited opportunity to exercise control, and role conflict. However, the hypothesis tested in this study, that stress occurs more in lower level jobs, was not completely supported. Yet increased stress among blue-collar workers has been reported by other researchers. Schwalbe and Staples (7) found a relationship between lower occupational position and stress: workers with routine work and least control over jobs could expect to experience the highest levels of stress. Schwalbe and Staples argued that their findings support other studies "that have documented the deleterious health effects of a labour process that denies autonomy and control, minimises intellectual challenge, creates surplus stress . . . of a whole class of direct producers" (7, p. 599). Otto (134) has also shown from a study of different occupational groups that managers report the lowest levels of stress while factory workers are aware of the greatest stress. She found factory workers had the highest stress scores of all groups studied, and Caplan and associates (83) found them to be the most disadvantaged. Kohn and associates (133) also found in the United States and Japan that factory workers report more distress than those in higher occupational groups. Andries and coworkers (102) have also argued that the impact of stress on blue-collar workers in Europe should not be underestimated.

That blue-collar women reported more stress than blue-collar men indicates differences in needs to be satisfied at work, and possibly different types of work. Braverman (2) and Hill (244) argue that women occupy jobs that have already been deskilled and fragmented and vacated by men. Game and Pringle (302) also argue that women suffer several disadvantages on the shopfloor. Women are given little comparative recognition for skills at work (they are often referred to as "dexterous" whereas men are mostly referred to as "skilled"). Also women experience a strong impact from the system of masculine control at the production plant level, one which male factory workers are more likely to accept and respond to. Chapter 4 outlines arguments that men in factory work tend to "pass off" the effects of negative work, and this may mitigate the already high stress effects for men. As the male work role is very important for identity, men may ignore some more stressful implications of factory work. No marked differences occur, however, between the bottom three-quarters of stress scores for women and men for each occupational group. While some studies have shown gender to be an important moderating variable (310), Guppy and Rick (311) also found no gender differences among British public service white-collar workers.

Special considerations for women at work have been argued by some writers. For example, Pattison and Gross (312) found that pregnancy could be difficult for women in already stressful jobs. Also Gross and associates (313) found that female corrections officers were more likely to be stressed than males. Greenglass (314) issues a caution, however, that previous research has not accounted for the social and structural aspects of gender in stress.

Psychological research reviewed in Chapter 2 shows that people see stressors differently and are stressed to different degrees by these experiences. A relatively small body of research has shown that structural location influences the extent to which objective phenomena are experienced as stressors (see 6, 7, 67).

Negative work experience was found to have differential effects on different occupational groups in the study sample. A lack of consideration, poor relationships, and role conflict represent negative aspects of management control of structure and of the climate of attitudes and values. Workers reported a stressfulness that is highly congruent with the amount of alienation they experienced. For the total sample, employees who experienced high alienation were likely to be stressed by that work characteristic, and those who experienced low alienation were likely to experience low stress.

"Lack of consideration by management" represents the climate of attitudes by management toward employees based on management's position of control in the organization. A greater lack of consideration by management is considered to be more stressful for all occupational groups. "Poor work relationships" and "role conflict," two aspects of negative work that result from management control of structure, also have high correlations with stressfulness for all occupational groups. Workers in each occupational group experienced important needs not being met due to poor work relationships and role conflict. There is a high congruence between stressfulness and the amount of alienation experienced. Dekker and Barling (315) found that role conflict, among other stress outcomes, could be related to large organizations.

The small amount of stressfulness reported by supervisors for the work factors "lack of skill discretion" and "freedom to control" indicates that they may accept that their role is difficult and that opportunities for skill use and varied tasks are somewhat limited. Factory workers reported a low level of skill discretion yet relatively low stressfulness. In Chapters 3 and 4 I argued that work routine is usually high in factory work due to job fragmentation and the severe division of labor resulting from highly mechanized work procedures. For a number of decades factory work has lacked skill content. Through socialization (past learning experiences), workers learn not to expect rewards such as work variety and the opportunity to use and develop skills in their work. Research into coping strategies (316) shows that many workers tend to lower their expectations about the possibility for need satisfaction at work, especially if they have not experienced these conditions in previous work. They may use coping modes to adjust to the lack of opportunity for important need satisfaction.

Managers have a large freedom to control and reported a small level of stress. However, there was a strong relationship between stress and a lack of control for factory workers. Again, through past socialization and work experience they are likely not to expect to satisfy needs for freedom to control at work. Gutterman and Jayaratne (317) also argue for the relationship between worker control and stress. In Chapter 3 I reviewed previous research (see 239) which showed that factory workers sometimes establish their own systems of exercising control and often "make do" at work in order to satisfy needs for control and influence that are not gained through formal work structures.

The relatively low stressfulness of high workloads for managers indicates some acceptance and successful coping approaches by them. Part of the status is acceptance of high workloads as a sign of valued contributions and work. The job is perceived as important, and a manager gains a sense of competence (as well as promotion and other rewards) in getting work done. For supervisors, a low association between workload and stress also indicates coping ability and acceptance as part of the job. A sense of self-worth and competence from getting work done may outweigh the deleterious effects of overwork.

Many stress studies that examine negative factors only at the level of the task have found task-related negative work to have an influence in determining stress response (65, 66, 109). In this study the task factors alone were studied and found to have an impact, but when considered with other elements of management control, they did not have a significant effect. Other studies (e.g., 83) have shown task factors to have a less strong impact when considered along with other negative work factors.

The effects of deskilling and of job fragmentation (a lack of skill discretion) and of perceived inability to influence how work is done (a lack of freedom to control work) may, on their own, be job elements that do not satisfy important needs. However, these effects are overridden by management control of structure and climate. For white-collar workers, each element has an effect on stress; for blue-collar workers, the effects of control of structure and climate are the most powerful predictors of the experience of stress. During group discussions prior to the study questionnaire being completed, survey participants referred to management's and supervisors' lack of appreciation and to specific characteristics of management style and practice as their greatest concern. While the opportunity to exercise skill discretion and control over the job may be important, many, especially lower level blue-collar workers, may have learned through past experiences that such opportunity in the work environment is negligible. The work of Otto (18), Menaghan and Merves (316), and others shows that employees may reduce their expectations when they have few positive work experiences and do not perceive their job as capable of providing such experiences—for example, the opportunity for control and skill utilization. The results of the current study show that outcomes of control over structure and climate of management attitudes are, for blue-collar workers, major causes of stress.

White-collar women showed no association between a lack of consideration by management and stressfulness. It is likely that as part of their culture they have accepted a certain degree of lack of consideration in a male-dominated management hierarchy. For male managers, the effects of management control over structure and a lack of consideration by management have important associations with stress. These may affect an important set of needs requiring satisfaction— needs likely to be associated with success in the organization and the male work role. For female managers there is no relationship between work factors and stressfulness.

Task factors have important effects on stress for women but not for men, and this has implications for the focus of some labor process studies. A number of these studies have found task variables to have important effects on stress (e.g., 4, 5, 7). While they are important, this study shows broader outcomes of management control over organizational structure and the climate of management attitudes to have overridingly important effects. However, for women there are associations between all levels of management control and stressfulness. This indicates the need for more research into gender issues, particularly at the level of task.

Perceived lack of control has important effects for women. I have outlined Game and Pringle's (302) arguments that women at work face a number of different forms of control based on masculine authority. They mainly occupy lower level blue- and white-collar positions and operate in these jobs under a masculine-based system of management control. For women, few important needs for using discretion and influencing the way work is done are met, supporting the argument that women may find a need for control hard to satisfy.

Those with the support of a partner who works are likely to find poor work relationships and the effects of role conflict to be more alienating than those without working partners or with no partner at all. Partnership support at home may give a more comparative basis on which to assess the quality and need fulfillment of poor work relationships and of conflicting role expectations.

CONCLUSION

A great deal of research into occupational stress (reviewed in Chapter 2) has shown that negative work experience is a major cause of stress. Labor process writers (e.g., 4–7) have shown that occupational stress is greatest for those in lower occupational groups, and argue that this results from working conditions in which workers lack control over the labor process.

Braverman (2) identified the causes of negative and alienating work as structural: he focused his argument about the effects of management control on the labor process on point-of-production activities. He argued that deskilling and job fragmentation have taken place. My earlier review of labor process research into occupational stress showed that researchers have focused on task-related activities as a primary cause of stress at work. The effect of these activities on stress in this study confirmed findings by Schwalbe and Staples (7). Littler and Salaman (162, 242) suggest that management control has brought about substantial increases in their control over structure and climate. Bureaucratic control has increased, in addition to management control of actual job activities.

The literature review in Chapters 2 to 5 showed that, in many studies, management's control of point-of-production activities has been found to contribute to the experience of stress. I argued previously that a lack of control or job discretion and a lack of task variety are two work characteristics that result from the separation between conception of the job by management and execution of the job by workers under management's direction. These characteristics also result from management's fragmentation of workers' tasks through division of labor and specialization, in most cases achieved through the use of mechanized work processes. However, even though important when analyzed on their own, the rest of the analysis in this chapter shows that these factors have little effect when considered together with other elements of management control over work.

The effects of three outcomes of management control were tested: control over task activities, control over structure, and the effects of control on the climate of attitudes and values of management toward workers. For the sample, the effect of management control of structure and the climate of attitudes and values are the strongest determinants of occupational stress. Task-related aspects such as control over job, skill utilization, and variety of tasks (skill discretion) have only a very small effect when included with aspects of control over structure and climate. This supports the argument that other effects of management control beyond the task level have important effects on stress.

Results support work by writers on stress in the labor process tradition. Management control of the labor process leads to work conditions conducive to stress. While point-of-production conditions are important to workers in not meeting needs and in creating stress, the effects of management control of structure and the climate of negative attitudes and values by management are the major determinants of stress. Child (161) argues that managerial strategic planning creates a structure and climate supportive of management's goals. While point-of-production activities are important, management control of structure and accompanying negative attitudes in the manufacturing environment in this study create the greatest alienation leading to occupational stress.

CHAPTER 9

Self-Esteem and Personal Mastery
at Work

This chapter considers two important sociopsychological consequences of negative work experience resulting from management's control of labor process organization: self-esteem and sense of personal mastery. Particular attention is paid to the relative effects of task-related variables, the outcomes of control over structure, and the climate of negative attitudes by management.

Low self-esteem and low sense of personal mastery are two intervening variables which, together with stress, are treated in the causal model presented in Chapter 6. Together with stress, low esteem and low mastery are the components of the sociopsychological response that mediate between noxious work and ill-health. We would expect to see differences in self-esteem and mastery among different occupational groups and between women and men. In addition, the effects of labor process organization on women's and men's self-esteem and sense of mastery are deemed to be important. In Chapter 4 I reviewed research which showed that negative work has important consequences for self-esteem and personal mastery. In studies of ill-health at work, the relationship between stress, self-esteem, and personal mastery is not always clear.

As described in Chapter 4, Schwalbe and Staples (7) argued that positive appraisals, favorable social comparisons, and the opportunity to perform work that develops a sense of competence affect levels of self-esteem. The ability to exert influence and use discretion in the workplace affects a being in exercising control over external events (sense of control or mastery). Blue-collar work is performed under conditions that are primarily controlled and defined by management, with little opportunity for workers to exercise influence and discretion. The impact of management control over work strengthens existing class differences through the experience of alienating work and associated negative sociopsychological consequences. Management control of structural arrangements at work affects the climate of attitudes and values, and this is likely to have important outcomes for employees' self-esteem and personal mastery.

179

SELF-ESTEEM AT WORK

Staples and associates argue that (176, p. 91):

> in the case of the perceived evaluation of others, or reflected appraisals, self-esteem arises from imagining that we are positively evaluated by others ... self-esteem deriving from social comparison is based upon the notion that we often arrive at judgements about self worth by comparing ourselves to referent others.

They maintain that self-efficacy is an important source of self-esteem and "the constraints imposed by the necessities of capitalist production also limit the worker's potential to experience self-efficacy, and may thus have detrimental consequences for self-esteem" (176, p. 92).

According to Schwalbe and Staples (7), an understanding of control of the labor process can help in locating low self-esteem as an important outcome of a capitalist mode of production (7, p. 588):

> Under capitalist relations of production, deriving self-esteem from efficacious action is difficult for those in working-class positions precisely because the capitalist labour process denies them autonomy, control and opportunities to demonstrate competence in challenging work.

Occupational and Gender Differences in Self-Esteem

Occupational and gender differences show the effects of social structure on self-esteem. Table 36 reports scores for low self-esteem for each of the five major occupational groups in the study. Both managers (16.61) and skilled workers (22.76) score below the sample average: this represents relatively high self-esteem. The lowest esteem levels are experienced by factory workers (27.87). Supervisors (23.54) and white-collar office workers (24.30) have mean scores somewhere between managers and factory workers. Major differences existed between factory workers and managers only (SNK test). White-collar workers have the greatest distribution in esteem scores.

There is little occupational stress literature on gender differences in self-esteem. In the organization in this study, women experienced marginally lower esteem than men as evidenced by the small but insignificant gender differences between white-collar (office) women and men.

Table 37 shows percentiles for low self-esteem. The bottom quarter of managers have self-esteem scores (3.3 or below) more than five times lower than those of the bottom quarter of factory workers (16.7 or below); this group of managers recorded very high esteem. Large differences exist for both 25th percentile scores and median scores between women and men in the skilled and

Table 36

Low self-esteem: mean scores by occupation and gender[a]

	Mean (S.D.)	n
Managers	16.61 (13.79)	56
Supervisors	23.54 (11.39)	49
White-collar (office)	24.30 (17.34)	45
Skilled	22.76 (14.94)	29
Factory	27.87 (16.85)	47
Total	22.77 (15.39)	226

$F = 3.86; p < .01$

	Women		Men		
	Mean (S.D.)	n	Mean (S.D.)	n	t
Managers	12.96 (11.95)	9	17.30 (14.12)	47	−.86
Supervisors	24.24 (13.77)	22	22.96 (10.43)	27	.37
White-collar (office)	27.31 (19.41)	31	17.62 (8.91)	14	1.78
Skilled	35.83 (14.24)	4	20.67 (14.21)	25	1.98
Factory	30.00 (15.42)	31	23.75 (19.61)	16	1.21
Total	26.49 (16.67)	97	19.97 (13.77)	129	3.25*

[a]Differences in numbers for the whole sample and the sum of each of the occupational and gender groups for the low self-esteem (10-item) scale are due to missing cases. The t values are for gender differences.
*p < .05

factory-work groups. Women in these groups have a much lower level of esteem than men.

Major Determinants of Low Self-Esteem

A regression model was developed to test factors that contribute to self-esteem. The causal model presented in Chapter 6 shows low self-esteem (together with stress and personal mastery) as an outcome of negative work experience, which in turn is treated as an outcome of the influence of ascribed and achieved statuses.

The total effects for three separately entered blocks of variables on low self-esteem are presented in Table 38. The first block includes ascribed and achieved statuses; the second, occupational position and length of time in the one position; and the final block, the effects of negative work. The measure of self-esteem used is Rosenberg's Inventory, an additive 10-item scale, scored from 0 to 100 (see Chapter 6).

Ascribed and achieved status variables explain 7% of variation in low esteem scores (adjusted R-squared = .07). Total variance explained by the model is 17%.

Table 37

Percentile table of low self-esteem by occupation and gender

	25th Percentile	50th Percentile	75th Percentile	(n)
Managers	3.3	16.7	26.7	(56)
Supervisors	15.0	23.3	33.3	(49)
White-collar (office)	11.7	20.0	33.3	(45)
Skilled	11.7	23.3	33.3	(29)
Factory	16.7	30.0	36.7	(47)
Total	10.0	23.3	33.3	(226)

	Women				Men			
	25th	50th	75th	(n)	25th	50th	75th	(n)
Managers	0.0	16.7	25.0	(9)	3.3	16.7	30.0	(47)
Supervisors	13.3	28.3	33.3	(22)	16.7	23.3	30.0	(27)
White-collar (office)	13.3	26.7	40.0	(31)	10.0	18.3	23.3	(14)
Skilled	26.7	30.0	50.8	(4)	10.0	20.0	31.7	(25)
Factory	16.7	30.0	36.7	(31)	6.7	21.7	35.8	(16)
Total	15.0	26.7	36.7	(97)	10.0	20.0	30.0	(129)

Therefore, only a fairly small amount of variability in self-esteem is explained by factors in the model. The greatest change in variability occurs when negative work experience is added.

Gender differences and years of education are the only status variables to have some effect on self-esteem. Differences between women and men (beta = −.13) have a small effect only. Compared with secondary-educated employees, those with only primary education are likely to have 7 points lower self-esteem; they have nearly 13 points less esteem than apprentice-trained workers (Figure 7). Tertiary-educated employees are likely to score 15 points more self-esteem than primary-educated employees. Differences between apprentice-trained, tertiary-educated, and secondary-educated employees are moderate.

The second hypothesis tested is that employees at lower occupational levels are expected to have lower self-esteem. As Schwalbe and Staples (7) showed, work that offers little opportunity for control and is characterized by job routine is most likely to occur in lower occupational positions. This is likely to lead to low self-esteem. Other studies support the effects of negative work experience such as lack of control and job routine on poor self-esteem. Schwalbe and Staples argue that "class positions that permit substantial autonomy control and opportunities to succeed in challenging work can facilitate the formation of efficiency-based

Table 38

Determinants of low self-esteem: regression coefficients

Independent variables	Dependent variable, low self-esteem	
	b	beta
Ascribed and achieved statuses		
Age	−1.00	.07
Primary education	7.38	.08
Apprenticeship	−5.49	−.12
Tertiary education	−7.67	−.23*
No partner	3.13	.09
Partner not working	3.56	.0
Gender	−4.15	−.13
Number of children	−0.28	−.02
Adjusted R-squared		**.07**
Occupational status		
Managers	−3.11	−.09
White-collar (office) workers	−1.27	−.03
Skilled workers	1.22	.03
Factory workers	3.01	.08
Years in one position	−0.47	−.04
Adjusted R-squared		**.06**
Negative work factors		
Role conflict	0.33	.02
Poor relationships	0.77	.03
Lack of freedom to control	−1.78	−.11
No skill discretion	5.31	.22*
Overwork	−4.15	−.18*
Lack of consideration by management	6.19	.27*
Adjusted R-squared		**.17**

*Significant at < .05

self-esteem" (7, p. 588). As Figure 8 shows, differences between factory workers and managers are moderate. A factory worker is likely to have over 6 points less self-esteem than managers. As occupational status diminishes, so does self-esteem.

"Lack of opportunity to use skill discretion on the job" has a moderate (beta = .22) effect. "Overwork" is the final negative work factor to have an effect: its

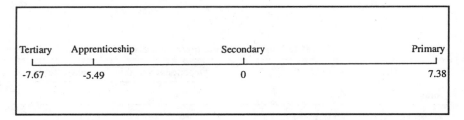

Figure 7. Unstandardized regression coefficient (*b*) scores: education and low self-esteem.

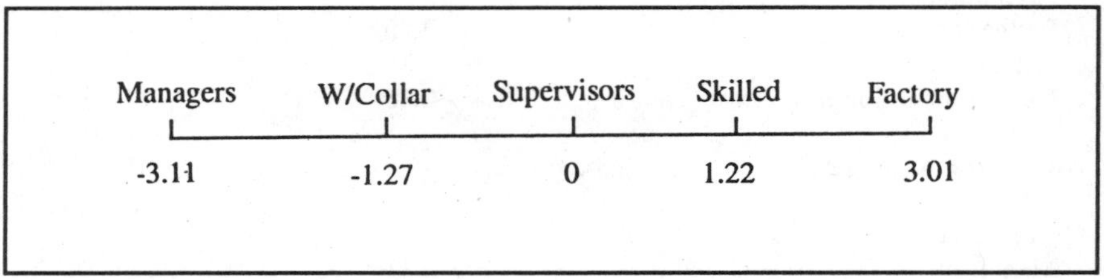

Figure 8. Unstandardized regression coefficient (*b*) scores: occupation and low self-esteem.

effect is small but significant (–.18). The greater the perceived workload, the higher the esteem. Negative work factors related to management control of structure (role conflict and poor relationships) have no real effect on levels of self-esteem.

Determinants of Low Self-Esteem for Occupational Groups

In order to test the effects of different factors on white-collar and blue-collar workers, separate regression equations were calculated for managers, supervisors, and white-collar (office) workers and for blue-collar skilled and unskilled workers. Only status variables from the first two blocks that have an effect significant at <.05 for either of the groups are shown (Table 39). The effects of all negative work factors for each group are included.

The *R*-squared for white-collar workers shows that all variables in the model explain 20% of variability in low self-esteem scores. Negative work factors alone account for 18% of the variation in scores. For blue-collar workers, very little variation in low esteem scores (*R*-squared = .04) is explained by factors in the model. For white-collar groups, moderate differences exist in the effects of

Table 39

Determinants of low self-esteem by occupational group

Independent variables	Dependent variable, low self-esteem (beta)	
	White-collar	Blue-collar
Primary education	.09	.08
Apprenticeship	−.13	−.10
Tertiary education	−.23*	−.06
Negative work factors		
Poor relationships	.06	−.14
Role conflict	−.01	.04
Overwork	−.13	−.08
No skill discretion	.27*	.15
Lack of freedom to control	−.02	−.23
Lack of consideration	.25*	.26
Adjusted R-squared	**.20**	**.04**

*Significant at < .05

education; having some tertiary education is likely to lead to increases in self-esteem. Education level provides no advantage for blue-collar groups.

Women and men were tested separately to determine the different factors affecting low esteem (Table 40). For women, 30% of variability in esteem scores is explained by the model: this represents a moderate to large influence on variations in self-esteem scores. Only 4% of variation in men's esteem scores is explained by the model. Tertiary-educated women are more likely to experience higher self-esteem than women with less education. Lower occupational positions are likely to lead to lower levels of self-esteem for women. "Overwork" has a high negative effect on low esteem for women, meaning that little work leads to low esteem. "Lack of freedom to control work" has a negative effect, and "lack of consideration by management" and "role conflict" have a low to moderate effect on women's self-esteem. Only "lack of consideration by management" has a moderate effect on men's self-esteem levels.

Discussion

Schwalbe and Staples (7) are some of the few authors who link the experience of low self-esteem at work to issues of broader class relations. They argue that management control of the labor process leads to disadvantageous work experience for lower class workers. This leads to the development of a poor sense of self-worth.

Table 40

Determinants of low self-esteem by gender

Independent variables	Dependent variable, low self-esteem (beta)	
	Women	Men
Tertiary education	−.27*	−.19
Occupational position	.30*	−.06
Negative work factors		
Role conflict	.25*	−.03
Poor relationships	.10	.13
Overwork	−.45*	−.09
No skill discretion	.19	.13
Lack of freedom to control	−.37*	.03
Lack of consideration from management	.30*	.30*
Adjusted *R*-squared	**.30**	**.04**

*Significant at < .05

As I discussed in Chapters 3 and 4, writers such as Braverman (2) and Thompson (238) have argued that management's control of the labor process has led to alienating working conditions. In this study, managers experienced the least negative working conditions and the highest self-esteem. Conversely, factory workers reported the greatest alienating work experience together with the poorest levels of esteem.

"Lack of consideration" is the only work factor to have an effect on men's self-esteem. For men, the work role is usually considered the primary and most important role in life. Men tend to be socialized into accepting many negative features of work life. While a lack of appreciation by management can affect a sense of being valued and therefore self-worth, many other negative features of work are likely to be accepted as part of the male hierarchy of control. Self-worth and competence and favorable social comparison are gained through other means. A lack of opportunity for skill discretion and a lack of consideration by management are the only alienating work factors to have a large effect on esteem scores for white-collar workers.

No negative work factors have a significant effect on low self-esteem levels for blue-collar workers. "Lack of consideration" has the highest effect (beta = .26). For white-collar workers, work experience contributes to low self-esteem to a small to moderate degree.

The effects of low self-esteem at work are considered further in the final Discussion section, following the report of findings on personal mastery.

PERSONAL CONTROL AND MASTERY AT WORK

In Chapter 3 I showed that the sociopsychological concept of mastery has been used in a number of studies on the effects of stress on ill-health outcomes. Some studies employing the concept of mastery variously assign it a mediating role in relation to stress outcomes (see Chapter 4). Pearlin and Schooler argue that a sense of mastery or control is one of a number of resources "that people draw upon to help them withstand threats posed by events and objects in the environment" (99, p. 5). Israel and associates (101) further argue that mastery is related to a sense of being in control of one's life.

A review of the literature has shown that few studies link sociopsychological concepts such as mastery to the wider class structure. Studies of stress and ill-health have paid little attention to the effects of the capitalist labor process on a sense of mastery. Based on labor process arguments on the effect of management's increased control over the labor process, the greater the impact of management control, typically at the blue-collar level, the less likely a strong sense of personal control over external events can develop.

The opportunity to exert influence and discretion over work and the work environment in order to achieve certain need satisfaction can influence the level of mastery or personal control. Negative work experience, resulting from management's increased control over point-of-production activities, structure, and the climate of attitudes and values, may lead to or substantiate an already existing low sense of personal control.

The study included a series of tests to examine the experience of low mastery through differences in occupational position and gender. The sense of mastery scale is an additive seven-point scale scored from 0 to 100 points (see Chapter 6). It measures the degree to which one perceives control over events in the world compared with being fatalistically ruled. Each of the seven items of mastery comprising the scale has a four-point response scale. Prior to score conversion, the response category of 0 means strongly agree; 1, agree; 2, disagree; and 3, strongly disagree. Therefore a high score represents a low sense of mastery. A number of studies (e.g., 101, 178) have used various measures of mastery.

Occupational and Gender Differences in Sense of Mastery

Relatively little research has been conducted on the effects of occupational position on mastery levels. Based on a small amount of literature presented in Chapter 4, lower occupational groups can be expected to have lower levels of mastery.

As Table 41 shows, scores range from a low mean score for managers (20.5) to a high score for factory workers (36.2). Occupational differences in mastery are large. Managers and white-collar (office) workers have scores below the sample average (29.3), while supervisors and skilled and factory blue-collar groups are

Table 41

Lack of personal control (mastery): mean scores by occupation and gender[a]

	Mean (S.D.)	n
Managers	20.5 (14.7)	55
Supervisors	30.5 (13.8)	50
White-collar (office)	29.2 (17.2)	44
Skilled	32.8 (13.4)	29
Factory	36.2 (15.1)	48
Total	29.3 (15.9)	226

$F = 7.77; p < .01$

	Women		Men		
	Mean (S.D.)	n	Mean (S.D.)	n	t
Managers	15.87 (14.29)	9	21.43 (14.81)	46	−1.03
Supervisors	30.74 (14.30)	22	30.27 (13.73)	28	.12
White-collar (office)	31.75 (19.06)	30	23.81 (10.89)	14	1.45
Skilled	41.67 (16.67)	4	31.43 (12.67)	25	1.44
Factory	37.79 (14.23)	31	33.33 (16.58)	17	.98
Total	32.39 (16.90)	96	27.07 (14.68)	130	2.29*

[a]Differences in numbers for the whole sample and the sum of each of the occupational and gender groups for the seven-item mastery scale are due to missing cases.
*$p < .05$

above the mean. Major differences exist between managers and all other groups (SNK test).

No real differences exist between women and men in the same occupational groups. Differences for skilled workers are largest, but there are insufficient numbers of women in that group. The bottom quarter of managers have a very low mean score for a lack of mastery (9.5) compared with all other occupational groups (Table 42). Their median score (19.0) is also substantially lower than that for all other groups. Fifty percent of managers experience a very high level of mastery. The only difference between gender groups is the lower score for the bottom quarter and median for female versus male managers. In the bottom quarter of factory workers, men have a higher degree of mastery than women.

Determinants of a Lack of Mastery

Table 43 shows the total effects for a number of variables on poor sense of mastery. Ascribed and achieved statuses were entered first, followed by occupational status then negative work experience. Eleven percent of variability in low

Table 42

Percentile table of low control (mastery) by occupation and gender

	25th Percentile	50th Percentile	75th Percentile	(n)
Managers	9.5	19.0	33.3	(55)
Supervisors	22.6	33.3	38.1	(50)
White-collar (office)	19.0	28.6	36.9	(44)
Skilled	26.2	33.3	40.5	(29)
Factory	25.0	33.3	47.6	(48)
Total	15.5	33.3	38.1	(226)

	Women				Men			
	25th	50th	75th	(n)	25th	50th	75th	(n)
Managers	2.4	14.3	31.0	(9)	9.5	19.0	34.5	(46)
Supervisors	19.0	31.0	39.3	(22)	25.0	33.3	38.1	(28)
White-collar (office)	17.9	33.3	38.1	(30)	17.9	23.8	33.3	(14)
Skilled	33.3	33.3	58.3	(4)	23.8	33.3	40.5	(25)
Factory	33.3	33.3	47.6	(31)	19.0	33.3	45.2	(17)
Total	20.2	33.3	41.7	(96)	14.3	28.6	38.1	(130)

mastery scores is explained by ascribed and achieved status variables. The total variance explained by the model is 22%, a moderate explanation of variations in scores. More than three-quarters of variance in scores is explained by variables outside the model. Differences in the effects of the number of years of education are greatest between those with a tertiary education and those with a secondary education. Having a tertiary education is likely to lead to over 11 points of increased mastery compared with a secondary education (Figure 9). Differences in mastery between tertiary-educated employees and both in primary-educated and apprentice-trained employees are also large.

Differences between having no partner and a working partner are moderate. Having no partner is likely to lead to almost 6 points more of low mastery. Figure 10 shows a moderate decrease in mastery for no partnership and a small decrease for having a nonworking partner.

Figure 11 shows large differences in mastery between some occupational groups. Factory workers experience almost 11 points lower mastery than managers. Differences between skilled workers and managers are less (nearly 8 points out of 100). From high to low occupational levels, level of mastery decreases.

Table 43

Determinants of a lack of mastery

Independent variables	Dependent variable, lack of mastery	
	b	beta
Ascribed and achieved statuses		
Age	0.26	.02
Primary education	−2.60	−.03
Apprenticeship	−2.57	−.06
Tertiary education	−11.55	−.33*
No partner	5.78	.16*
Partner not working	2.14	.05
Gender	−2.77	−.09
Number of children	1.98	.13
Adjusted R-squared		**.11**
Occupational status		
Managers	−5.84	−.16
White-collar (office) workers	−0.41	−.01
Skilled workers	2.93	.06
Factory workers	5.02	.13
Years in the one position	−0.48	−.04
Adjusted R-squared		**.13**
Alienating work		
Role conflict	−0.15	−.01
Poor relationships	1.70	.07
Overwork	−2.07	−.09
No skill discretion	3.03	.13
Lack of freedom to control	−0.23	−.01
Lack of appreciation from management	7.46	.32*
Adjusted R-squared		**.22**

*Significant at < .05

Only one factor, "lack of appreciation by management," has a large effect on mastery (beta = .32). "Lack of skill discretion" has a small effect (.13). Other work factors have a minor effect only. Contrary to expectations, task-related activities ("lack of freedom to control" and "lack of skill discretion") have no effect on workers' sense of mastery; that is, demonstrating competence and exerting influence over the job do not affect employees' sense of personal control.

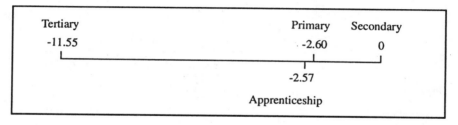

Figure 9. Unstandardized regression coefficient (*b*) scores: education and lack of personal mastery.

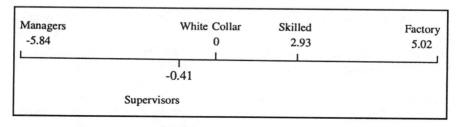

Figure 10. Unstandardized regression coefficient (*b*) scores: partnership status and lack of personal mastery.

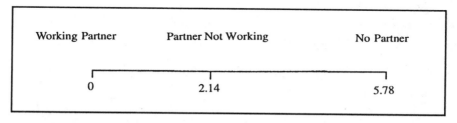

Figure 11. Unstandardized regression coefficient (*b*) scores: occupation and lack of personal mastery.

Effects of Occupational Position on Sense of Mastery

In Chapter 4 I argued that those in lower class positions are likely to perceive little control over the external world. In order to test the influence of structural location on beliefs about control over external events, two separate regression equations were calculated for white-collar workers (managers, supervisors, and white-collar (office) workers) and blue-collar workers (Table 44). Total effects of

Table 44

Determinants of lack of mastery by occupational group

	Dependent variable, lack of mastery (beta)	
Independent variables	White-collar	Blue-collar
Tertiary education	−.36*	−.14
No partner	.09	.33*
Number of children	.05	.27*
Negative work factors		
Poor relationships	.03	.12
Role conflict	.03	−.15
Overwork	−.15	.12
No skill discretion	.10	.10
Lack of freedom to control	.03	.06
Lack of consideration	.28*	.20
Adjusted *R*-squared	**.21**	**.07**

*Significant at < .05

variables are given for each group. Only those ascribed and achieved status variables with a moderate to large effect are shown, together with all negative work factors.

An R-squared of .21 for white-collar workers shows that a moderate amount of variance in mastery scores is explained. Only 7% of variance for blue-collar mastery is explained by the variables in the model. The R-squared values for ascribed and achieved status variables plus years in the one position are similar (.08 for white-collar workers; .07 for factory workers). However, adding negative work factors for white-collar workers produces an increase of 13% in variance explained. For factory workers, the added factors do not add to the explained variance: work factors alone do not explain variations in their scores.

Large differences in low sense of mastery exist between tertiary and other education levels for white-collar workers: the greatest differences are between those with primary versus secondary levels of education. There are no real advantages for tertiary-educated employees in blue-collar occupations.

For blue-collar workers, having no partner (beta = .33) is likely to be a large disadvantage compared with having a working partner, and more disadvantageous than having a nonworking partner (−.12). Another aspect of family life that has an effect on mastery for blue-collar workers is family size: the more children, the more likely that resources to deal with threats from the environment (mastery) can be strained (.27). "Lack of consideration by management" is the only negative work factor to have a moderate influence on white-collar workers' sense of mastery. For blue-collar workers, "lack of consideration" has only a small effect on mastery, but the greatest effect of all work factors.

Gender Differences in Sense of Mastery

Little research has been carried out on the different determinants of a lack of personal control in the workplace. Two separate regression equations were calculated for men and women (Table 45). For women, 29% of variation in mastery levels is explained by factors in the model. However, for men, only 16% of variance is explained. For both women and men, having a tertiary education leads to a greater internal locus of control (beta = −.36 for women, −.32 for men). Also, as occupational level decreases, the likelihood of having control over events decreases for both women and men (.31 and .21, respectively). "Lack of consideration by management" is an important determinant of lack of mastery for both women and men. For women, "poor work relationships" also has a moderate effect on low mastery. Women reported that "overwork" (−.27) has a small but insignificant effect on sense of mastery, while men reported no effect.

DISCUSSION

As Chapter 7 showed, white-collar office workers have the lowest level of negative work, followed by managers and supervisors. Managers, however, are

Table 45

Determinants of lack of mastery by gender

Independent variables	Dependent variable, lack of mastery (beta)	
	Women	Men
Tertiary education	−.36*	−.32*
Occupational position	.31*	.21*
Negative work factors		
Role conflict	.13	−.05
Poor relationships	.24*	−.13
Overwork	−.27	.00
No skill discretion	.09	.05
Lack of freedom to control	−.22	.10
Lack of consideration from management	.34*	.35*
Adjusted *R*-squared	**.29**	**.16**

*Significant at < .05

likely to experience the greatest levels of self-esteem due to their position of influence and control. Favorable reflected appraisals are high for managers, as shown by low scores on a "lack of consideration" by management. Chapter 7 also showed that managers experience very little (.72) lack of skill discretion compared with factory workers (1.79). Favorable social comparisons are likely to be low wherever there is little opportunity for skill use and variety. Social comparisons are likely to be positive for managers due to their position of control and opportunity to use their prerogative of management. Sense of competence is likely to be high: managers exert a large degree of influence, use skills freely, and have very little routine in their work.

As we more down the occupational hierarchy, workloads become less and esteem levels drop. Some workers, particularly managers, gain a sense of being needed when workloads are high. An increased workload can indicate to employees their value to the workgroup or the organization. A sense of competence and worth is likely to be higher for managers. In lower level jobs where workers' values are not so closely tied to organizational goals, a high workload can represent an excessive demand.

When management gives negative appraisals to employees, such as not treating them with respect, not appreciating their work, not listening to their ideas, and not telling them they are doing a good job, a poor sense of self is a likely outcome. The less positive feedback from management and the less management shows respect, the lower the self-esteem of workers. The effects of management control are strongest on the factory floor. Child (161) and Burawoy (239) substantiate that blue-collar workers are likely to experience the most negative working environment due to the effects of management control. However, I have shown that it is not only point-of-production (task) activities (as Braverman argued) that have an effect: workload demands and the effects of a negative climate of management attitudes are also important determinants of a low sense of worth.

Large differences occurred in the study for "lack of consideration" scores with low and moderate levels of poor esteem for managers and white-collar workers (results not shown). For these groups, a low "lack of consideration" was associated with poor esteem scores compared with greatly increased lack of consideration associated with moderate self-esteem. Other white-collar groups (supervision and white-collar office workers) tend to have middle-class values supportive of management's position of control. Poor reflected appraisals gained from management are likely to affect a sense of worth. A lack of consideration reflects management's attitudes about the value of employees, their contribution and worth to the organization. White-collar groups are likely to be most susceptible to the influence of these appraisals. A lack of consideration, as argued in earlier chapters, derives from management's position of control and belief in the prerogative of management.

Schwalbe and Staples (7) also argued that negatively reflected appraisals and poor social comparisons gained from working in jobs with fewer opportunities and reduced task opportunities contribute to low self-esteem for working-class (blue-collar) workers. Factory workers and skilled blue-collar workers (to a smaller extent) have very little influence and few opportunities to use discretion over their tasks, giving limited opportunities for developing competence at work. Work process and structure are controlled by management, offering few opportunities for favorable social comparisons. A large degree of negative attitudes from management offers little in the way of positive reflected appraisals.

Based on expectations from research reviewed earlier, the major hypothesis tested in this chapter, that "as occupational position decreases, self-esteem levels will diminish," was supported.

There is a good deal of variation in self-esteem scores for white-collar workers, possibly due to the variation in jobs from more routine typing and secretarial positions to more responsible and varied accounting and computer work. Managers are expected to display middle-class values and work with high levels of influence and control. All levels of managers usually support the values of the organization (growth, productivity, and competitiveness) and are normally regarded as valued workers. In most cases they are able to gain high esteem through performing tasks that are considered important in the organization.

Skilled workers' esteem levels are better than for all levels except managers. A sense of competence gained from craft membership and from having highly skilled jobs is likely to give them a positive sense of self, even though they report fairly large negative attitudes (reflected appraisals) from management. The low scores of blue-collar workers reflect their performing work that can often be replaced by mechanized processes and tends to have low value placed on it. Financial and other extrinsic rewards are normally low, and these workers rarely get the opportunity to use skills that can give feelings of competence and self-worth on the job.

On the basis of literature reviewed in Chapter 4, the hypothesis that self-esteem is expected to decrease with lower occupational position was tested and verified. Managers reported the highest esteem, and factory workers the lowest. When workers have little opportunity to use previously learned skills or to learn new ones on the job, the opportunity to demonstrate competence on the job is limited. A lack of variety in tasks also limits a sense of competence, particularly in completing a whole job. Each of these aspects is present to a large degree in factor work and to a lesser degree in the work of white-collar groups.

Of the negative work factors, "lack of consideration from management" is a primary determinant of low self-esteem (beta = .27). Management control over elements of task activities, job demand requirements, and the climate of attitudes and values leads to diminished self-esteem for many workers; however, lack of consideration by management has an overriding effect.

More years of education are likely to lead to greater promotion opportunities and more mobility for white-collar employees, and their training may make them more suited to gaining management positions. Higher self-esteem would reflect these greater opportunities and the more responsible positions that tertiary-educated employees are likely to occupy.

As shown in Chapter 7, white-collar (office) workers experience a lack of opportunity for skill discretion to a moderate degree. Supervisors experience a lesser lack of opportunity, and managers a small lack of skill discretion. White-collar workers achieved a higher level of education in the sample than blue-collar workers and would expect to employ this training at work. Managers' and supervisors' jobs have less deskilling due to technological change than do lower level white-collar (office) jobs. In fact, even lower level management positions have increased their dimension of control by taking more of the planning function out of lower level jobs. As I showed earlier, white-collar (office) work, according to Braverman (2) and other labor process writers, has undergone quite a degree of deskilling. White-collar workers would normally expect to have the opportunity to use their skills and to work at a range of tasks. Their sense of competence is likely to be affected by increased routine in work and little skill use and development. Less favorable social comparisons are likely as jobs may appear less useful and less important than those which give employees opportunities to fully utilize their skills and resources.

Women's self-esteem is influenced by a wide variety of negative work factors. Control over task level, the level of management control of structure, and the negative climate of attitudes all have an effect on women's low esteem. For men, very little in the model explains low self-esteem: a negative climate of attitudes is the main determinant of any effect. As I have shown earlier, writers such as Game and Pringle (302) argue that women at work experience the negative outcomes not only of a management-controlled labor process but of a male-controlled labor process. Lower self-worth develops from unfavorable social comparisons of men's greater opportunity for promotion and other rewards and their doing work considered to have greater value than women's work.

The work role is one of a number of important roles in life for women. They derive much of their identity from domestic and mothering roles. Work experience for women is often an opportunity to develop a sense of competence and self-reflected appraisals outside their domestic and family role, and the experience of alienating working conditions greatly affects their sense of self. In this study, women's sense of self-esteem is lower than men's. Game and Pringle (302) argue that women's work roles, being more limited than men's, traditionally deny them opportunities for developing their competence.

For women, the advantage experienced through education results in higher occupational status that is more likely to provide positive appraisals from management and give better opportunities for a sense of competence. For women in more senior positions, having achieved promotion and higher status in areas mostly occupied by men is likely to lead to very favorable social comparisons and development of high levels of competence. Women in lower occupational groups, however, have lower self-esteem. Factory work and white-collar (office) jobs for women are often deskilled. Factory floor jobs are one part of a male work sphere. This supports Hill's (244), Thompson's (238), and Game and Pringle's (302) arguments presented earlier that women often fill the least rewarding work roles.

Having less work to do leads to a diminished self-concept. Low workload is likely to be associated with jobs that are considered "not all that important." These employees are likely to receive negative reflected appraisals. Compared with home roles where they have principal control over domestic and child-rearing responsibilities, a low workload is likely to be associated with not being valued. Factory women report the lowest workloads. Their work typically develops little sense of competence and poor social comparisons. In addition, receiving conflicting orders and demands can bring about a feeling of powerlessness and therefore lower personal worth.

The low contribution of work experience to self-esteem scores indicates that low levels of self-worth are already established through working-class experience for blue-collar workers. Socializations that develop low expectations and limited competence together with unfavorable social comparisons and poor reflected appraisals are likely to be strongly established. Work experience is likely to confirm a class-based learning of a sense of self-worth. Previous work experience often substantiates this. White-collar workers are normally socialized into fairly high expectations from work, which are likely to be affected quite strongly by work experience.

The results of variations in self-esteem through work experience support the Marxist position that the working-class experience of alienating work leads to a diminished sense of self. Important needs of the individual (self-esteem and worth) are likely not to be met for blue-collar workers: they are also more likely to be influenced through work experience for women than for men.

Lower scores on mastery for factory workers compared with managers indicate that workers in lower class occupational positions (blue-collar skilled and factory workers) are more prone to believe they have little chance to influence external events. White-collar (office) workers are also likely to perceive little opportunity to affect changes. Tertiary-educated employees experience more mastery and occupy management positions. Their opportunities for promotion are usually greater than for those with less education. This can lead to a greater belief in the ability to influence external events. As shown in Chapter 7, managers have the greatest opportunities at work through the use of discretion, control, and influence to bring about changes in their work conditions. Correspondingly, they are likely to have a strong belief in their ability to solve problems and to change situations (internal locus of control). Blue-collar (factory) workers experience least control and opportunity to use discretion and affect changes at work.

Having a partnership increases mastery. Those with increased support and a higher income base at home are likely to effect more changes in their life than those without the equivalent support. Gender differences are very small, with women likely to report slightly lower levels of mastery.

A lack of feedback from management, a lack of appreciation for work by the supervisor and management, and a lack of respect are all aspects of the climate of attitudes and values derived from management's control and belief in the prerogative of management. These attitudes negate the value of employees' contributions. They express a particular aspect of alienation that denies the value of the person: labor produced is the most valued item. These negative attitudes by management create a sense of powerlessness: employees may feel that it is not possible to exert influence because management shows no real interest in or appreciation for their contributions. A lack of consideration and appreciation demonstrates a work environment of dominance by management.

Mastery levels are more likely to be formed and established outside the work experience (previous socializations and previous work experience). Factory work and skilled work, which are quite high in negative work content, are likely to confirm previously held beliefs in an ability to exert less influence than in higher level jobs. For white-collar workers, work experience is likely to confirm established beliefs in a greater personal control.

Blue-collar work offers fewer opportunities for promotion and general mobility than white-collar work. Increased economic responsibility can reduce the ability to deal with situations positively and constructively as they arise. Having a working partner and dual incomes can provide greater emotional support, allowing workers to make more financial and career-related choices and hence have a greater internal locus of control.

A greater lack of consideration is likely to lead to low levels of mastery for white-collar workers. Middle-class values including high levels of education are likely to lead managers, supervisors, and office workers to value being appreciated in their work. If their role is not valued, they are likely to feel that their base for influence and for affecting control as a manager is not strong. In Chapter 4 I reviewed literature which shows that a need for achievement and an ability to deal effectively with problems are important for managers.

Neither blue- nor white-collar workers' mastery levels are affected by control over task-related activities or by the effects of control on structure in this study. Only white-collar workers experienced a negative climate of attitudes as a determinant of low mastery. The results indicate that a great deal of lack of mastery for blue-collar workers is established outside the work experience.

In their domestic and mothering roles, women are likely to have established a sense of mastery through family and schooling socializations. As shown in previous chapters, Hill (244), Braverman (2), and Baxter and coworkers (306) argue that women have fewer opportunities for employment, promotion, and mobility than men. Work experience and biographical variables explain more about women's mastery levels than men's. This inequality is likely to affect mastery levels. Game and Pringle (302) argue that work experience is based on a male-based system of control that denies women many traditional opportunities for exercising a sense of control. Work experience may also contrast with domestic roles in which women can exert influence. Work for men is a taken-for-granted role; many of the negative experiences at work are less likely to strongly influence their sense of control.

For women, poor work relationships influence their level of mastery. Such relationships are most characteristic of jobs where technology and workflow requirements determine workgroup relations. Relationships with others are one of the benefits for many women in leaving the domestic role. A supportive workgroup can encourage and develop a sense of problem-solving and effectiveness in being able to bring about changes.

Having control over task-related factors has no influence on either women's or men's sense of personal mastery levels.

CONCLUSION

Pearlin and coworkers (178) argue that a sense of mastery is a resource that allows people to withstand threats from the environment. A negative work environment can lead to low self-esteem and low mastery. Labor process writers have argued that working-class (blue-collar) workers experience the most negative work conditions (e.g., 2, 7). These can include conditions related to the work itself, or to the effects of management-controlled structure and the climate of attitudes and values that result from management control. Where work experience is controlled mostly by management, alienating working conditions are likely to occur. Based on the work of Pearlin and Schooler (99), Pearlin and coworkers (178), others, alienating work experience is conducive to low self-esteem and low sense of mastery.

A sense of mastery can develop from positive work experiences. As a socio-psychological state, it can produce a positive sense of being able to deal with problems and effect positive changes for the future. In interaction with the other two sociopsychological variables—stress and self-esteem—it can help to draw on resources such as coping strategies which can negate the effects of negative or harmful influences in the work environment. As was expected from past studies reviewed earlier, negative work experience explains less about the formation of low personal mastery than it does about the experience of the other sociopsychological variables, stress and low self-esteem.

Little literature is available on the effects of sociostructural factors on the experience of mastery. Large occupational differences were found in the experience of a lack of mastery. As occupational level decreases, level of personal control decreases. Blue-collar workers who experience the most alienating work also experience the least mastery.

Schwalbe and Staples (7) tested for the effects of two task-related factors, "perceived lack of control" and "job routine," on low self-esteem. The results of this study (not reported) confirm their findings on the effects of "job routine" on low self-esteem. The effects of task-related variables together with other elements of management's control were tested on both low self-esteem and low sense of mastery. When included with other outcomes of management's control, task-related factors have very little effect.

Braverman (2) argued that the impact of management's control has been greatest at the level of the task. However, in this study both task-related factors and a lack of consideration by management, together with workload, were important determinants of low self-esteem. The effects of a negative climate of management attitudes toward workers had the greatest impact on low levels of mastery. The effects of management's control of the labor process through task variables had a minor effect only in reducing mastery.

The ability of white-collar employees (managers, supervisors, and white-collar (office) workers) to deal with problems in the work environment (lack of mastery) was affected most by a climate of negative attitudes and by having little work to do. This is consistent with middle-class values placed on acceptance and on productive and meaningful work. Simile to low self-esteem, negative work had only a small effect on blue-collar workers' abilities to deal with problems. Home factors such as having no partnership support and/or having large numbers of children had the greatest effects on mastery for blue-collar workers, indicating that economic resources may have some role in being able to deal with problems in the work environment. Class differences can affect the ability to withstand external threats. As with self-esteem, it is white-collar workers' and women's sense of mastery that is most affected by work experience.

These findings are in contrast to the focus of much labor process research on stress and ill-health, which has directed attention at point-of-production activities as the basis of alienating work experiences. This has not been confirmed for self-esteem and mastery, which require a broader conceptualization of management control to understand its effects.

CHAPTER 10

Effects of Negative Work Experience and Stress on Health

Considerable research has been conducted on the effects of various types of working conditions on ill-health (see, e.g., 104, 318 for reviews of some of this research), and research into physical ill-health has shown occupational stress to be a determinant (see Chapter 2). Lesser but still positive associations have been demonstrated between other sociopsychological states such as low self-esteem and low sense of personal mastery and ill-health.

The aim of this chapter is to examine the extent of ill-health symptom frequency for different occupational groups and for women and men. It will also examine the major determinants of somatic and emotional ill-health and the effects of sociostructural factors (occupation and gender) on ill-health.

Based on the review of the literature, the model of stress and ill-health described in Chapter 6 shows ill-health (somatic and emotional) as the outcome of a whole series of causal relationships. It shows that ascribed and achieved statuses, occupational position, and length of time in the one job all have effects on sociopsychological states. These relationships were tested and presented in Chapters 8 and 9.

SYMPTOM AWARENESS:
OCCUPATIONAL AND GENDER DIFFERENCES

The symptom awareness scale used in this study is a compilation of emotional and somatic symptoms constructed from symptom checklists used in a number of different studies (refer to Chapter 6 for the various sources). The 27-item symptom checklist has been reduced to 25 items for scaling purposes. Two questions, on menstrual problems and trouble caused by menstrual periods at work, were not included because of large numbers of missing cases. The scale is additive and has been scored from 0 to 100. Prior to converting the scale, the response categories for each symptom item were 0, never or hardly ever; 1, sometimes; and 2, almost always or always.

There are large variations in the types of health scales used in studies of stress. Margolis and Kroes (179) suggest that certain stress-related health outcomes can be measured to give an idea of the effects of stress on workers. First, there are short-term outcomes from specific job stresses such as anger, tension, and anxiety. Second, longer term psychological responses such as chronic depression can become a part of the individual's health profile rather than being a general response to specific conditions. Third, clinical-physiological changes, as in blood pressure and blood lipids, are objective measures of stress responses. The authors argue that these outcomes can lead to psychosomatic or physical illness. Finally, symptoms such as gastrointestinal disorders and coronary heart disease are obvious physical or psychosomatic responses.

Mean scores for the different occupational groups were compared (Table 46). Large occupational differences are evident in the awareness of physical and emotional symptoms. Substantial differences exist between managers (mean, 18.00 out of 100) and factory workers (33.16). Mean scores for supervisors, skilled workers, and white-collar workers are below the sample average and are well below the mean for factory workers. There were significant differences in symptom frequency reporting between factory workers and all other groups (SNK test).

Table 46

Ill-health symptom awareness: mean scores by occupation and gender[a]

	Mean (S.D.)	n			
Managers	18.00 (13.17)	53			
Supervisors	21.48 (14.22)	46			
White-collar (office)	23.15 (14.79)	40			
Skilled	22.59 (17.07)	27			
Factory	33.16 (17.63)	43			
Total	23.46 (15.98)	209			

$F = 6.29; p < .01$

	Women		Men		
	Mean (S.D.)	n	Mean (S.D.)	n	t
Managers	22.00 (15.57)	8	17.29 (12.77)	45	.93
Supervisors	28.33 (15.59)	18	17.07 (11.51)	28	2.82*
White-collar (office)	26.44 (14.93)	27	16.31 (12.38)	13	2.12*
Skilled	33.50 (14.82)	4	20.70 (17.00)	13	1.41
Factory	39.19 (17.64)	27	23.00 (12.46)	16	3.22*
Total	30.30 (16.91)	84	18.46 (13.20)	125	5.74*
	$F = 3.07$; n.s.		$F = .87$; n.s.		

[a]Differences in numbers for the whole sample and the sum of each of the occupational and gender groups are due to missing cases. The t values are for gender differences.
*$p < .05$

In order to test for other sociostructural differences in the awareness of symptoms, women and men in each occupational category were compared. Moderate to large differences in symptom awareness frequency are evident between female and male supervisors, white-collar (office) workers, and factory workers. In all cases, women experienced a greater symptom frequency.

Table 47 shows percentile scores of ill-health for occupational and gender groups. The 25th percentile score for factory workers is twice that of all other groups except office workers. Their median score (32.0) is twice that of managers and supervisors: thus the bottom half of factory workers reported a much higher symptom awareness than the bottom half of managers and supervisors. The 25th percentile scores for female white-collar (office), skilled, and factory workers are more than double the scores for the equivalent groups of men; that is, the quarter of women reporting the least symptoms for each of these groups reported a far greater symptom awareness than men. Median score differences between women and men for supervisors, skilled workers, and factory workers are also large.

FREQUENCY OF SPECIFIC SYMPTOMS: OCCUPATIONAL AND GENDER DIFFERENCES

In order to test the effects of sociostructural factors on specific groups of symptoms, occupational and gender differences were tested. Several studies have

Table 47

Percentile table of ill-health symptom awareness by occupation and gender

	25th Percentile	50th Percentile	75th Percentile	(n)
Managers	6.0	16.0	28.0	(53)
Supervisors	10.0	16.0	32.5	(46)
White-collar (office)	14.0	22.0	33.0	(40)
Skilled	10.0	20.0	30.0	(27)
Factory	20.0	32.0	44.0	(43)
Total	11.0	22.0	33.0	(209)

	Women				Men			
	25th	50th	75th	(n)	25th	50th	75th	(n)
Managers	6.0	24.0	34.5	(8)	6.0	16.0	27.0	(45)
Supervisors	10.0	29.0	42.0	(18)	10.0	14.0	19.5	(28)
White-collar (office)	14.0	22.0	40.0	(27)	5.0	18.0	26.0	(13)
Skilled	10.0	37.0	45.5	(4)	8.0	20.0	24.0	(23)
Factory	20.0	42.0	46.0	(27)	13.0	22.0	32.0	(16)
Total	11.0	30.0	43.5	(84)	8.0	16.0	26.0	(125)

reported occupational differences in the awareness of emotional and somatic symptoms: some show differences in the frequency of perceived symptoms, while others show differences in presenting complaints to a medical practitioner.

The factor analysis presented in Chapter 6 identified six factors from the ill-health symptoms checklist. Each factor is composed of a number of individual symptoms and represents a specific symptom awareness group. Mean (average) scores are presented for occupational groups and for women and men within occupational groups. Standard deviations are given in parentheses, representing the degree to which symptom frequency varies within each group. The F scores indicate differences in the manufacturing environment in the study; they are shown both for occupation and for gender within occupations. Mean scores for each factor are on a scale of 0–2, with 0 meaning hardly ever or never; 1, sometimes; and 2, always or nearly always experiencing the symptoms.

Emotional Problems. This is a compilation of items relating to mental and emotional health and well-being. Aspects of mental and emotional well-being including depression, anxiety, and irritability have been used in a variety of studies. Graham states that "there have been a number of studies examining the role of various stresses in the aetiology of neurosis and increased psychological symptomatology" (319, p. 154). Karasek (65, 66), Pearlin and associates (178), Caplan and coworkers (83), and others use a number of indices including depression, and/or anxiety, and/or irritability.

There are moderate occupational differences in reported emotional symptoms (Table 48). Managers (.58) reported the least symptom awareness. Their work is high in control and skill discretion, and they report the most positive experience from other effects of management control of the labor process. Significant differences exist between factory workers and managers only (SNK test). Karasek (66) found least depression resulting from work experience when workers had sufficient latitude to exercise control and discretion. Work with the greatest mental ill-health consequences occurred when an underutilization of opportunities for discretion on the job was combined with heavy task demands. Supervisors and skilled and semi-skilled blue-collar workers reported similar levels of emotional problems in this study (.68), but more than managers (.61). Even these levels are quite high for reported emotional and mental ill-health symptoms: they are more than half way between "hardly ever" and "sometimes" experiencing all seven symptoms, on average. Supervisors have a fairly large degree of control over their own work and the work of others. They reported a fairly low degree of alienating work experience. Skilled and semi-skilled blue-collar workers reported the next highest degree of alienating work compared with factory workers. However, skilled workers reported a level of stress comparable to the low levels reported by managers. It is likely that membership and identification with strongly based craft and skill groups could obviate some of the harmful effects of negative work experience.

Table 48

Emotional problems: mean scores by occupation and gender

	Mean (S.D.)	n
Managers	.58 (.44)	54
Supervisors	.68 (.44)	48
White-collar (office)	.74 (.51)	44
Skilled	.61 (.52)	28
Factory	.87 (.46)	45
Total	.69 (.47)	219

$F = 2.71; p < .01$

	Women		Men		
	Mean (S.D.)	n	Mean (S.D.)	n	t
Managers	.73 (.60)	8	.55 (.41)	46	1.09
Supervisors	.87 (.51)	20	.54 (.33)	28	2.75*
White-collar (office)	.88 (.52)	30	.43 (.32)	14	3.00*
Skilled	1.07 (.50)	4	.53 (.49)	24	2.05
Factory	1.03 (.46)	28	.60 (.33)	17	3.41*
Total	.92 (.50)	90	.54 (.39)	129	6.08*

*$p < .05$

White-collar (office) workers reported a higher level of emotional symptoms than managers, supervisors, and skilled workers. For this group, work demand is moderate (overwork), as is the opportunity for skill discretion and control. The higher score occurs despite their reporting low levels of negative work experience. The results for factory workers are consistent with the findings of Caplan and coworkers (83) and a number of other reports on emotional disorders. While factory workers reported almost 50% more of these symptoms than managers, the differences between occupational groups are only moderate.

Musculoskeletal Disorders. This group of disorders includes pains and injuries related to the spine. Back and neck pain is frequent in certain types of occupations. Where lifting is involved, awkward movements can result from workplace design and workflow requirements. Postural discomfort results from long spells of sitting or standing or from specific body movements required by the technology which can lead to musculoskeletal symptoms. Strain or discomfort—as well as accidents—are likely to cause such symptoms. Corlett argues that "many of the diseases of middle age are contributed to in a major way by posture inadequates" and that machine operators are not the only workers to experience bad posture:

"the requirement to stand all day set by many jobs is unequivocally a bad posture. So too is the requirement to sit all day" (320, p. 41).

Differences between occupational groups are moderate to large (Table 49). Factory workers reported these symptoms most frequently (.99): more than twice the frequency for managers (.43) and supervisors (.49). Their score indicates often experiencing musculoskeletal problems, on average. In the manufacturing organization in the study, unskilled factory work was based around machine processing, with each worker a member of a team working one machine. The machine process generally regulates the movements of team members. Some of their work entails lifting with possible awkward movements due to machine-process demands. Machine pacing also largely determines the workplace of the group.

White-collar workers reported the next highest occurrence of back and neck pain and related problems. Most office work is done sitting during the work day. VDU screens and other office technologies such as keyboards are likely to create some ill-health: postural and strain problems are common. Skilled blue-collar workers and supervisors reported musculoskeletal symptom awareness below the sample average (.59 and .49, respectively). Major differences exist between factory workers and other groups (SNK test).

Table 49

Musculoskeletal disorders: mean scores by occupation and gender

	Mean (S.D.)	n
Managers	.43 (.48)	54
Supervisors	.49 (.51)	50
White-collar (office)	.66 (.57)	45
Skilled	.59 (.51)	28
Factory	.99 (.57)	47
Total	.63 (.56)	224

$F = 8.50; p < .01$

	Women		Men		
	Mean (S.D.)	n	Mean (S.D.)	n	t
Managers	.47 (.53)	8	.42 (.48)	46	.24
Supervisors	.59 (.63)	22	.40 (.39)	28	1.30
White-collar (office)	.74 (.58)	31	.46 (.53)	14	1.54
Skilled	.50 (.33)	4	.60 (.54)	24	−.37
Factory	1.16 (.57)	30	.71 (.44)	17	2.83*
Total	.81 (.62)	95	.49 (.48)	129	4.28*

*$p < .05$

Gastrointestinal Problems. Individual symptoms of stomach problems, diarrhea, and nausea make up "gastrointestinal problems." These can be a related depressive response, but the somatic symptoms can also result from overstimulation or understimulation of production of important hormones such as adrenaline and noradrenaline. As Table 50 shows, there is very little occurrence of gastrointestinal symptoms (mean .18) in the study sample. All groups except factory workers reported close to the mean score. Factory workers reported an occurrence of these symptoms (.31) more than twice that of managers (.13), white-collar (office) workers (.14), and skilled blue-collar workers (.14). Major differences are between factory workers and all groups except skilled workers (SNK test).

Otto has argued that digestive disturbances are a common symptom of prolonged stress and that "there is a risk of duodenal or stomach ulcers relating to continued over-secretion of gastric acid and impairment of protective stomach lining" (18, p. 52), with high blood pressure and hypertension. These symptoms may also be related to infectious conditions, but not necessarily.

Allergies and Allergic Reactions. According to Frank Austen (291), "allergies," in particular allergic rhinitis, is a group of symptoms commonly referred to

Table 50

Gastrointestinal symptoms: mean scores by occupation and gender

	Mean (S.D.)	n			
Managers	.13 (.24)	55			
Supervisors	.18 (.31)	49			
White-collar (office)	.14 (.26)	43			
Skilled	.14 (.29)	28			
Factory	.31 (.44)	49			
Total	.18 (.32)	224			

$F = 2.69; p < .05$

	Women		Men		
	Mean (S.D.)	n	Mean (S.D.)	n	t
Managers	.21 (.31)	8	.12 (.22)	47	.97
Supervisors	.21 (.25)	21	.15 (.36)	28	.57
White-collar (office)	.14 (.29)	30	.13 (.22)	13	.18
Skilled	.08 (.17)	4	.15 (.31)	24	−.43
Factory	.36 (.47)	32	.22 (.37)	17	1.14
Total	.24 (.34)	95	.15 (.29)	129	1.95

*$p < .05$

as "hay fever." It is related to hypersensitivity. He argues that "perennial allergic rhinitis occurs in response to . . . the processed materials or chemicals utilised in industrial settings or the dust accumulating at work or at home (291, p. 1412). Other types of allergies may be implied or associated and are characterized by skin and throat problems. According to Austen, increased emotional pressure may increase allergic susceptibility. Also Graham notes that "there is growing evidence from a number of prospective and longitudinal studies that stress increases susceptibility to acute respiratory infections" (319, p. 154). The cellular immune response has been shown to be influenced by psychological factors and stress.

The frequency of allergic symptoms reported by factory workers in the study (.57) was more than double that of managers (.24) (Table 51). Supervisors and white-collar workers have scores equivalent to or below the study sample mean (.40). Skilled workers also reported a fairly high frequency of these symptoms (.51). Occupational differences are small to moderate. Major differences exist between managers and factory and skilled workers (SNK test).

In the manufacturing organization in the study, blue-collar work is carried out in exposed locations on the factory floor, centered around processing machines. Products being processed contain large amounts of dust and airborne substances

Table 51

Allergies and allergic reactions: mean scores by occupation and gender

	Mean (S.D.)	n
Managers	.24 (.32)	55
Supervisors	.35 (.36)	50
White-collar (office)	.40 (.46)	44
Skilled	.51 (.47)	28
Factory	.57 (.56)	47
Total	.40 (.45)	224

$F = 4.51$; significant at $< .01$

	Women		Men		
	Mean (S.D.)	n	Mean (S.D.)	n	t
Managers	.29 (.49)	8	.23 (.29)	47	.53
Supervisors	.48 (.43)	22	.24 (.24)	28	2.57*
White-collar (office)	.46 (.51)	30	.29 (.32)	13	1.14
Skilled	1.08 (.74)	4	.42 (.34)	24	3.01*
Factory	.66 (.54)	31	.42 (.58)	17	1.40
Total	.54 (.53)	95	.29 (.35)	129	4.05*

*$p < .05$

that are likely to lead to irritation and allergic infection. The whole factory design is open, with minimum protection for workers from airborne pollutants. Forklift drivers are constantly exposed to raw materials that are high in airborne contamination.

As argued in Chapter 1, cortisol plays an important role in the immune response. Increases in cortisol have been associated with situations over which people perceive they have little control. Otto argues that control "reduces inflammation and pain which is the body's response to injuries or to foreign substances (including bacteria and others)" (18, p. 50). Under prolonged stress, when cortisol secretions are important in sustaining antibodies to maintain some resistance, allergic conditions are likely to occur.

Analgesic Overuse and Related Disorders. Constipation, dizziness, and headaches are typical symptoms of the overuse of analgesics and pain-killing drugs such a codeine. As such, these symptoms are related to a number of other emotional and somatic symptoms. These symptoms can also result from an interaction with other conditions. As shown in Table 52, factory workers experience the greatest frequency of these symptoms (.56), twice that of skilled workers (.21)

Table 52

Symptoms of analgesic overuse and related functional disorders: mean scores by occupation and gender

	Mean (S.D.)	n
Managers	.22 (.32)	55
Supervisors	.37 (.38)	49
White-collar (office)	.38 (.34)	43
Skilled	.21 (.32)	27
Factory	.56 (.41)	48
Total	.36 (.38)	222

$F = 6.98$; significant at $< .01$

	Women		Men		
	Mean (S.D.)	n	Mean (S.D.)	n	t
Managers	.38 (.45)	8	.19 (.28)	47	1.54
Supervisors	.57 (.35)	21	.23 (.33)	28	3.54*
White-collar (office)	.46 (.34)	29	.21 (.28)	14	2.35*
Skilled	.58 (.57)	4	.14 (.22)	23	2.83*
Factory	.68 (.42)	31	.33 (.31)	17	2.97*
Total	.56 (.39)	93	.21 (.29)	129	7.34*

*$p < .05$

and managers (.22). Supervisors and white-collar (office) workers are close to the sample mean; moderate to large occupational differences exist in the frequency of headaches, dizziness, and constipation ($F = 6.98$). There are major differences between factory workers and all other groups, and between managers and supervisors (SNK test).

Factory workers reported a higher frequency of analgesic use than the other groups. Table 53 shows the differences in frequency of use by occupational group. Skilled blue-collar workers reported the least use of pain-killing drugs. These differences are likely related to experiencing other emotional symptoms that result from more negative work conditions, more stress, and lower levels of personal worth and mastery that would enable workers to deal positively and energetically with problems and a lack of need fulfillment. In this sense, the effects of analgesic use can be related to experiencing little influence over the labor process.

In every occupational group (except managers), women reported significantly more symptoms of analgesic overuse than men (trouble with breathing, skin problems, and sore throats and colds). Table 54 shows differences in frequencies of use by women and men in the study. The proportion of women (15%) taking analgesics more often than once per week was more than twice that of men (5%). And the proportion of women (34%) taking analgesics more often than once per month was twice that of men (17%). Gender disproportions are slightly higher than the national average reported in the Australian Health Survey (1989–90) (321): in the two weeks prior to the survey, 56.3% of women and 45.2% of men had taken pain relievers. Differences in the symptoms of analgesic overuse in this study are similar to those found during the 1983 Australian Health Survey (322) (37 per 1,000 women versus 15 per 1,000 men).

Functional Depressive Conditions. Two symptoms associated with more severe depressive reactions are appetite problems and difficulty getting to sleep or staying asleep (249). They are part of a group of functional emotional and

Table 53

Analgesic use by occupational group (percent in categories)

Frequency of use	Managers	Super-visors	White-collar (office)	Skilled	Factory	Total (n)
More than once per week	7	12	9	4	12	9 (21)
More than once per month	26	20	27	14	33	25 (56)
Less often or never	67	68	64	82	55	66 (149)
Total (n)	100 (55)	100 (50)	100 (44)	100 (28)	100 (49)	100 (226)

Table 54

Analgesics use by gender (percent in categories)

Frequency of use	Women	Men	Total
More than once per week	15	5	9
More than once per month	34	17	24
Less often or never	51	78	67
Total (n)	100 (100)	100 (131)	100 (231)

depressive disorders that are more likely to be associated with parasympathetic nervous system responses, where individuals' responses to conditions over which they have little control are passive. These conditions can be part of a problem of passively withdrawing from problems. Occupational differences are insignificant. Factory workers reported the greatest symptom frequency (.64), and white-collar (office) workers reported the lowest (.40) (Table 55).

DETERMINANTS OF SOMATIC AND EMOTIONAL SYMPTOM AWARENESS AND FREQUENCY

In Chapters 2 to 4 I reviewed research on the causes of occupational ill-health. Based on the causal model developed in the study (Chapter 6), the independent and mediating variables were tested to find the relative strength of their effects on ill-health. Table 56 shows the total effect of four blocks of independent variables. Ascribed and achieved statuses are entered together, followed by occupational status variables, negative work factors, and finally, sociopsychological outcomes. A factorial ANOVA model tested for independent variables with high correlations. No interaction effects were found for independent variables in the model.

Mechanic (67) argues that sociopsychological responses have an intervening role—that is, in adaptation to external demands and in mediating resulting ill-health consequences. Sociopsychological variables (stress, self-esteem, and mastery) are now treated as intervening variables in the model, outcomes of negative work experience that affect the awareness of emotional and somatic symptoms.

Forty-three percent of variations in ill-health scores are explained by all of the included variables (Table 56). Ascribed and achieved statuses explain a moderate amount of variability (15%). After occupational level and number of years in the one position are added, 23% of variance is explained; after including negative work factors, 29% is explained. Some variations in ill-health scores are explained by general class location factors (e.g., education and occupation). Finally, substantial variations in ill-health scores are explained by the mediating effects of stress, low self-esteem, and low sense of mastery.

Table 55

Functional depressive conditions: mean scores by occupation and gender

	Mean (S.D.)	n
Managers	.41 (.57)	55
Supervisors	.46 (.46)	48
White-collar (office)	.40 (.43)	45
Skilled	.53 (.58)	29
Factory	.64 (.55)	49
Total	.48 (.52)	226

$F = 1.82$; n.s.

	Women		Men		
	Mean (S.D.)	n	Mean (S.D.)	n	t
Managers	.63 (.74)	8	.37 (.53)	47	1.16
Supervisors	.58 (.52)	20	.36 (.40)	28	1.51
White-collar (office)	.42 (.45)	31	.36 (.41)	14	.44
Skilled	.75 (.50)	4	.50 (.60)	25	.79
Factory	.67 (.49)	32	.59 (.67)	17	.50
Total	.57 (.51)	95	.42 (.53)	131	2.01*

*$p < .05$

Educational level had a moderately large influence on the awareness of symptoms. Differences are large between tertiary-educated employees and every other educational group (Figure 12). A difference of nearly 11 points (out of 100) of ill-health symptom awareness is likely between tertiary-educated employees and those with only primary education. Differences between tertiary- and secondary-educated employees are moderate but less pronounced (more than 6.5 points).

The total effect of education on symptom awareness is $-.20$, indicating that educational advantage is likely to lead to ill-health. Little of the effects of education is unexplained by the model (direct effect .02). There is a large indirect component ($-.22$) in explaining the effects of education: a great deal of effect is through intervening variables (these figures are not shown in the regression model). Less than one-third of the indirect effects, however, are attributable to the effects of educational advantage on occupational position. The positive health benefits of higher education have an effect through more positive work experience and the effects of beneficial sociopsychological resources and outcomes.

Women are aware of more emotional and somatic symptoms than men (beta = $-.30$). A b score (unstandardized regression coefficient) of -19.32 shows that women are likely to experience over 19 points (out of 100) more of ill-health symptom awareness than men.

Table 56

Determinants of symptom awareness: regression coefficients

Independent variables	Dependent variable, symptom awareness	
	b	beta
Ascribed and achieved statuses		
Number of children	2.19	.15
Primary education	4.17	.04
Apprenticeship	1.30	.03
Tertiary education	6.66	−.19*
No partner	0.23	.01
Nonworking partner	3.22	−.08
Age	0.83	−.05
Gender	−19.66	−.30*
Adjusted R-squared		**.15**
Occupational status		
Managers	3.06	.08
White-collar (office) workers	0.61	−.02
Skilled workers	3.88	.08
Factory workers	5.34	.14
Years in one position	4.12	.31*
Adjusted R-squared		**.23**
Alienating work		
Role conflict	0.70	.03
Poor relationships	2.50	.11
Lack of freedom to control	0.38	−.02
Lack of skill discretion	2.08	.08
Overwork	0.15	−.01
Lack of consideration	5.87	.25*
Adjusted R-squared		**.29**
Sociopsychological responses		
Stress	11.38	.34*
Low self-esteem	4.43	.13
Low mastery or control	4.10	.12
Adjusted R-squared		**.43**

*Significant at < .05

Occupational status was the second block of variables entered into the regression equation. (Powles and Salzberg (323) note the lack of studies on occupational differences in health.) As shown in Figure 13, there are moderate differences for symptom awareness between different occupational positions. Factory workers experience more awareness of symptoms than white-collar (office) workers (more than 6 points out of 100). There is a smaller difference between factory workers and managers. Factory workers reported the greatest awareness of symptoms, followed by skilled and semi-skilled blue-collar workers.

Number of years in the one position has a significant effect on symptom awareness (beta = .31). Being in the same position longer and perceiving poorer health is partly explained by the large number of lower level workers who stay longest in the one job.

The Effects of Negative Work Factors

Only one negative work factor has a small to moderate effect on symptom awareness: a lack of consideration by management (beta = .25). An increase in each step of the four-point lack of consideration scale from "never" to "always" is

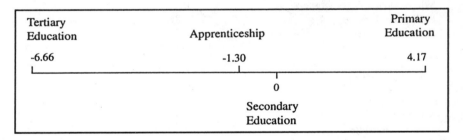

Figure 12. Unstandardized regression coefficients (b) scores: education and health.

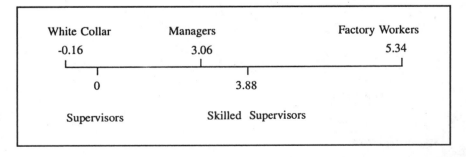

Figure 13. Unstandardized regression coefficients (b) scores: occupation and health.

likely to increase symptom awareness by nearly 6 points (out of 100). The difference between "never" and "always" a lack of consideration is about 18 points increase in symptom awareness.

The Effects of Sociopsychological Variables

Selye (9, 15, 21) established a link between the experience of stress and increased levels of ill-health (see Chapter 1). According to the psychological and sociological research reviewed earlier, when the demands of the work environment do not match important needs of employees, or when the resources available in the environment are insufficient, distress is likely to occur. Depending upon the strength and availability of resources and the effects of other interacting sociopsychological variables such as self-esteem and mastery, ill-health is a likely result. A considerable amount of research into occupational stress has shown causal links between the experience of occupational stress and ill-health.

As Table 56 shows, stress, as expected, has a large effect (beta = .34). Low self-esteem, when controlling for the effects of stress and a lack of mastery only, has a small effect on symptom awareness (.13). Low self-esteem, however, has a larger effect than a lack of mastery (.12).

SYMPTOM AWARENESS: THE RELATIONSHIPS WITH STRESS, SELF-ESTEEM, AND MASTERY

To test the level of symptom awareness for workers with different socio-psychological outcomes, symptom frequency was compared with low, medium, and high states of stress, low self-esteem, and low mastery. Symptom frequency is strongly associated with changes in levels of stress. As stress levels increase, so does symptom awareness. Changes in symptom awareness are more strongly associated with changes in levels of mastery than with changes in self-esteem levels. Ill-health symptoms are measured on a scale of 0 to 100.

Table 57 shows that the mean score for symptom awareness increases by almost 25.00, or one-quarter of the entire range, as stress increases from low to high levels. Employees with high stress experience all symptoms, on average, "sometimes." Differences are large. However, symptom awareness increases less dramatically as levels of self-esteem decrease. For employees with a moderate level of self-esteem, average symptom frequency is 29.19 (out of 100). There are large increases in symptom awareness related to decreases in mastery levels. The mean score for symptom frequency moves from 17.88 to 45.78 as levels of mastery decrease from high to low. Low stress, high self-esteem, and high mastery are related to a low symptom awareness.

Table 57

Ill-health symptom frequency by stress, low self-esteem, and low mastery

	Mean (S.D.)	(n)
Stress		
Low	17.22 (12.48)	95
Moderate	27.69 (15.25)	67
High	41.25 (21.18)	16
Total	23.31 (14.49)	178
$F = 23.70**$		
Self-esteem		
High	21.17 (14.38)	157
Moderate	29.19 (18.49)	57
Low	No one reported this low level of esteem.	
Total	23.31 (15.57)	214
$F = 11.10*$		
Mastery		
High	17.88 (13.56)	85
Moderate	25.60 (15.19)	120
Low	45.78 (22.03)	9
Total	23.38 (14.89)	214
$F = 17.31**$		

$*p < .05$
$**p < .01$

Occupational Differences in Determinants of Ill-Health Symptom Frequency

In order to test for differences in effects of occupational membership on ill-health symptoms, regression equations were calculated for white- and blue-collar workers. This procedure tests the effects of class bases on ill-health reporting. Table 58 shows differences between combined groups of white-collar and blue-collar workers in emotional and somatic symptom frequency Three separate regression equations were calculated for each occupational category; each regression equation repeats the procedures used in Table 56. Four separate blocks of variables were added, including ascribed and achieved status variables, occupational status, negative work factors, and sociopsychological consequences of negative work experience. For both white-collar and blue-collar occupational groups, women reported a greater frequency of symptoms (–.31 for ᵥ –.45 for blue-collar). For blue-collar workers, 17% of variance is ᴇ ascribed and achieved statuses; 11% is explained for white-collar workers. For blue-collar workers, spending a long time in the one job is a strong determinant; it does not have a significant effect for white-collar workers. Factory and skilled workers have relatively little mobility and spend more time in jobs with little opportunity to exercise influence at work.

"Lack of consideration by management" is the only negative work factor to have a significant effect on symptom awareness for blue-collar workers (beta = .27). This demonstrates the strength of this alienating effect of management's control. The effect of a lack of consideration is the highest of all negative work factors for white-collar workers (.18), yet not significant. "Lack of control" has no effect, and task factors have no effect.

The effects of stress are an important determinant of ill-health symptom frequency only for white-collar workers (beta = .38). Low self-esteem has a small (.18) but insignificant effect, while low mastery has no effect. Both stress and low self-esteem have small but insignificant effects on symptom frequency for blue-collar workers (.28 for both). Low sense of mastery for blue-collar workers has no effect at all. Overall, the model explains a moderate amount of variation in symptom frequency for white-collar workers (.32), but half of the variation for

Table 58

Determinants of ill-health by occupational group

| | Dependent variable, ill-health symptoms (beta) | |
Independent variables	White-collar	Blue-collar
Gender	.31*	–.45*
Years in the one position	.21	.42*
Negative work factors		
Overwork	.04	.04
Lack of freedom to control	–.05	.06
No skill discretion	–.07	.19
Poor relationships	.11	.16
Lack of consideration by management	.18	.27*
Role conflict	–.06	.07
Sociopsychological outcomes		
Stress	.38*	.28*
Low esteem	.18	.28
Adjusted *R*-squared	**.32**	**.51**

*Significant at < .05

blue-collar workers (.51): the model explains a lot about blue-collar symptom reporting.

Determinants of Women's and Men's Ill-Health Symptom Frequency

Separate analyses were carried out for women and men; Table 59 shows four blocks of variables in regression equations for each. Only sufficiently large standardized (beta) regression coefficients are presented. Having a tertiary compared with a secondary level of education (beta = −.27) is likely to lead to a lower symptom frequency for men; for women this has a lesser effect (−.11). Length of time in the one job has a large effect on symptom reporting for women (.55): the longer the time in the same job, the more likely that symptom awareness will increase.

"Lack of consideration by management" has a moderate effect on symptom frequency reporting for women (.30); for men it is not significant. The greatest effect of negative work factors for women, however, is for poor relationships at work (.36). For men this has no effect. "Poor relationships" and "lack of skill discretion" are important predictors of ill-health symptom reporting for women. A

Table 59

Determinants of ill-health by gender

	Dependent variable, ill-health symptoms (beta)	
Independent variables	Women	Men
Tertiary education	−.11	−.27*
Years in the one position	.55*	.20
Negative work factors		
Role conflict	.04	.09
Poor relationships	.36*	−.08
Lack of freedom to control	−.08	.04
No skill discretion	.22	−.09
Lack of consideration by management	.30*	.23
Overwork	−.02	.02
Sociopsychological outcomes		
Stress	.09	.43*
Low esteem	.28	.10
Low mastery	.14	.05
Adjusted R-squared	**.50**	**.21**

*$p < .05$

lack of skill discretion (skill underutilization and work routine) has a small but significant effect (.22). No negative work factors have more than a small effect on men's symptom reporting. A great deal more variation in women's symptom frequency is due to alienating work factors: an additional 23% of variation in symptom awareness is explained. No additional variation in men's symptom frequency scores is explained with the addition of alienating work experience.

The greatest effect on symptom frequency for men is the stress that they experience (beta = .43). Both low esteem and a low level of mastery have virtually no effect. For women, low esteem has a small but insignificant effect on symptom awareness, while low mastery and stress have no real effect.

In summary, for men a higher education is likely to lead to fewer reported symptoms, while stress is likely to lead to ill-health awareness. Women who stay in the one job longest and women who experience negative attitudes from management are most likely to report the greatest symptom frequency.

DETERMINANTS OF SPECIFIC SYMPTOM GROUPS

A regression analysis was undertaken for each of the six specific groups of symptoms identified in the factors analysis (Table 60) in order to determine which independent and mediating variables contribute an effect for each of the specific

Table 60

Determinants of major symptom groups[a] (beta values)

	Sociopsych.: Emotional	Chronic: Musc. skel.	G.I.	Acute: Allergy	Analg.	Sociopsych.: Funct. depress.
Tertiary education	−.10	−.18*	−.11	−.16*	−.12	−.19*
Gender	−.35*	−.21*	−.10	−.23*	−.38*	−.09
Age	−.04	.04	−.02	−.23*	−.03	−.13
Factory workers	.03	.23*	.14	.09	.06	.08
Years in the one position	.17	.32*	.10	.34*	.21*	.06
Role conflict	.05	.19*	.11	.02	.05	−.04
Lack of consideration by management	.35*	.00	.14	.10	.26*	.08
Poor relationships	.13	−.01	.05	.03	−.07	.21*
Stress	.35*	.26*	.33*	.20*	.08	.17
Adjusted R-squared	**.38**	**.28**	**.07**	**.24**	**.25**	**.13**

[a]Sociopsych. = sociopsychological symptoms; musc. skel. = musculoskeletal problems; G.I. = gastrointestinal problems; allergy = allergies and allergic reactions; analg. = symptoms of analgesics overuse; funct. depress. = functional depression symptoms.

symptom groups identified in Chapter 6. This tests the effects of all variables in the model described in Chapter 6. Four separate blocks of variables were entered. Only moderate to large effects of independent variables are given.

Women are most likely to have increased emotional (depressive, anxiety, and irritability) symptoms (beta = .35). A negative climate of attitudes based on management's control (a lack of consideration from management) is also likely to have a large effect in leading to emotional disorders (.35). Finally, stress, where there is an imbalance between demands in the work environment and individual's important needs, is likely to lead to a high frequency of emotional symptoms (.35).

Women are more likely to experience musculoskeletal disorders than men (beta = −.21). In addition, working on the factory floor in unskilled jobs has a small effect (.23) compared with higher occupational groups. The only alienating work characteristic likely to have a small effect on disorders related to the back and spine is role conflict. As with emotional problems, stress is a strong determinant (.26).

Stress has a large effect on the frequency of gastrointestinal problems (beta = .33) and a lesser effect on allergies and allergic reactions (.20). In fact, it is the only moderate to large determinant of gastrointestinal symptoms. Age and length of time in the one position both have fairly important effects on reporting allergies. In all, women, younger people, and those with least mobility in the organization are likely to experience the most frequent colds, skin problems, and breathing disorders.

The experience of stress is not an important determinant for the last two groups of symptoms. Women are likely to experience the greatest frequency of symptoms of analgesic overuse (beta = −.38). As with allergies and allergic reactions, longer time spent in the one position is also likely to lead to greater symptoms of analgesic overuse. A lack of consideration by management is the only negative work factor to have some effect on these symptoms. Functional depressive symptoms, which are characteristic disorders of a deeper seated depression (difficulty with getting to sleep and staying asleep, and poor appetite), are affected by two factors that do not have strong effects on any of the other symptom groups. Tertiary-educated employees (−.19) are likely to experience fewer such symptoms. Poor relationships at work (.21), an outcome of management control over structuring of activities, are likely to lead to stronger emotional and depressive symptoms.

Women are likely to experience a greater number of symptoms more frequently than men. However, being stressed at work is the most likely cause of the greatest symptom reporting. The climate of attitudes and values resulting from management control has the most frequent effect of all negative work factors. A large amount (38%) of variation in the frequency of emotional problems is explained by variables in the model. Slightly less is explained for musculoskeletal disorders (.28), symptoms of analgesic overuse (.25), and allergic symptoms (.24). Only small amounts of variation in functional depressive symptoms (.13) and gastro-intestinal disorders (.07) are explained by ascribed and achieved statuses, occupational status, alienating work factors, and sociopsychological outcomes included in the model. The literature reviewed earlier has shown that women report many ill-health conditions more frequently than men. The stress literature shows that stress is an important precursor of disease.

DISCUSSION

As noted in Chapter 4, Navarro (6), Schatzkin (8), and Garfield (5), along with a small group of other labor process writers, argue that employees' different health levels are outcomes of differential class experience at work, due largely to the influence of management's control over the labor process. Chapters 7 to 9 have shown blue-collar work to have the largest component of alienating work characteristics and the highest levels of stress, lowest self-esteem, and poorest sense of mastery. This chapter has reported findings on the highest symptom frequency for blue-collar workers.

Many of the differences between blue-collar workers, white-collar workers, and managers are related to socioeconomic indicators such as income and education. Mathers (235), for example, has shown health differences between different educational, occupational, and income groups in Australia.

Managers, the group with the lowest symptom frequency, have the most favorable experience of work. They experience relatively few disadvantages of alienating working conditions at all levels: the job itself, the organization of structure, and the climate of negative attitudes from management. They have the lowest stress levels and highest self-esteem and mastery of all groups. Skilled blue-collar workers have a slightly lower level of symptom awareness than white-collar workers, but slightly higher than supervisors.

Relatively few studies have investigated occupational differences in ill-health (see 323). Those studies that do, as noted earlier, show the lowest occupational levels to be most at risk. One of the earlier studies supported by the research presented in this book is that by Caplan and associates (93), who reported on three types of symptoms in their study of stress and ill-health: anxiety, depression, and irritability. They found that machine tenders reported the highest symptom levels, and small groups of other unskilled and skilled workers had high anxiety, depression, or irritability scores. Also, Schwalbe and Staples (7), using a sample size and occupational distribution similar to the study sample in this book, reported a mean symptom score for somatic and emotional symptoms of 15.07 (based on a range of 10–30; a low score represents poor health). Thus they found poorer health levels than those in this study. However, Schwalbe and Staples included few managers and claim to have overrepresented semi-autonomous employees. House and associates (307), Otto (134), and others have also reported disadvantage to blue-collar unskilled groups.

Lowe and Northcott argue that "while a number of studies compare distress among males and females generally, there have been few attempts to examine men and women within the same occupation" (146, p. 57). In their study of Canadian postal workers, women were found to have higher distress (as measured by depression, irritability, and psychophysiological complaints) than men. The higher symptom frequency for women is not necessarily strongly related to differences in stress or other sociopsychological states, nor due only to greater levels of alienating work. For example, in Chapter 7 I reported that only among factory workers do women report significantly more alienating work than men. In addition, no stress differences are evident for women and men within occupational groups.

As argued earlier, women tend to have a greater number of important roles outside the workplace (such as homemaker and mother) than men. These increased role strains could account for women's reporting greater symptoms. In addition, women's sex role socialization has taught them to rely more on their feelings and emotions than men, which may also account for some higher symptom reporting. However, factory workers report significantly more alienating conditions (see chapter 7), and this greater negative work experience may account for women's higher symptom awareness in this group.

Chapters 3 and 4 discussed labor process studies which showed that alienating working conditions can lead, through a process of stress, to ill-health. According to Schwalbe and Staples (7), control and routinization, two aspects of the capitalist labor process, are likely to produce stress and low self-esteem, which in turn will lead to ill-health. They found that routine and, to a lesser degree, a lack of control were likely to be experienced more by blue-collar workers. These two conditions affected stress and self-esteem, which in turn were found to positively affect scores on an index of somatic and emotional health.

However, Coburn found that "objective alienation" (measures of repetition and lack of autonomy, with jobs rated accordingly) correlated "at best only weakly with the indices of general well-being" (4, p. 46). "Subjective alienation," a self-reported evaluation of interest and challenge in the job, correlated highly with measures of well-being. "Perceived alienation," self-reported work characteristics, correlated less highly with well-being. Correlations between alienation and physical well-being, however, were not high. This illustrates an important point: the need to look beyond the task level in explaining the factors contributing to ill-health.

Otto (324) found automobile workers to have less control than government factory workers. They reported greater stress and more frequent psychological and physiological complaints. O'Brien and coworkers (144) found that underutilization of skills and abilities was related to reduced mental well-being and frustration. Anxiety and fatigue occurred most frequently in jobs with low levels of control. However Caplan and colleagues' (83) study was one of the few to test a large number of occupational groups for related health outcomes. They found that measures of emotional ill-health were most associated with assembly work. There is sufficient evidence to demonstrate the labor process arguments of Schwalbe and Staples (7), Coburn (4), Garfield (5), and Navarro (6) that conditions created by increased management control over the labor process disadvantage the health of blue-collar (factory) workers and other lower occupational groups.

Major studies on gender differences in ill-health in Australia have shown that women's morbidity is higher than men's. As noted by Lupton and Najman (325, p. 9):

> health and health-related behaviour [is] one of the most striking examples of the differences between men and women in contemporary Australian society. Official statistics, health surveys and medical/hospital records, consistently show higher mortality rates for men, but higher rates of morbidity and health services use for women.

The 1983 Australia Health Survey (322) reported large differences in symptom reporting between women and men: 57 per 1,000 women reported nerves, tension, depression, and related disorders in the two weeks prior to interview in the survey, compared with only 34 per 1,000 men. Lupton and Najman (325) report from unpublished tables in a 1987 Brisbane study that in a four-week period women experienced psychosocial symptoms more frequently than men: 26% of women reported worry or depression, which was reported by only 15% of men. Lowe and Northcott (146) found that female postal workers experienced greater psychophysiological symptoms and depression than men. I found few gender differences in irritability in this study. The greatest gender differences in emotional symptom reporting were among factory workers.

Other studies have also reported more disorders among women in the general population. In Lupton and Najman's (325) sample, 19% of women reported joint pain or swelling, compared with only 12% of men. Accident or injury, however, was a greater cause of problems for men than for women: 11% of men versus 5% of women.

Gender differences have been reported in other Australian studies. Gender differences were found in the 1983 Australian Health Survey (322) for coughs and sore throats. Lupton and Najman (325) found 31% of women in their sample to have experienced colds, sore throats, or coughs over a four-week period. Only 24% of men experienced these symptoms. In the 1983 Australian Health Survey (322) women reported twice the incidence of allergy reported by men.

Broom argues that "women apparently suffer from higher rates of most kinds of morbidity, and consult medical professionals more often than men" (326, pp. 122–123). She notes that 67% of women and 57% of men in the two weeks prior to interview for the 1983 Australian Health Survey had one or more illness conditions. Based on the Australian Bureau of Statistics results, Broom shows that "headaches and migraines, arthritis, and hypertension all strike females more often than males" (326, p. 123). Due to differences in role responsibilities and in socialization, women are likely to report symptoms more frequently than men (see 302, 306, 326).

In this study, gender differences in symptom reporting have been found within occupational groups across a number of different occupations. According to Broom (326, p. 124);

> women suffer a higher incidence of unpleasant and disabling (but not fatal) conditions such as headaches, migraines, varicose veins, urinary tract and genital disorders, 'minor' psychiatric conditions (anxiety and depression) and arthritis.

Hraba and associates (327) also found that men reported lower depression, hostility, and anxiety than women.

Lupton and Najman (325) cite differences in socialization and in role obligations between women and men as reasons for different symptom reporting. In addition, Lowe and Northcott argue that the "double burden of job and family may create additional stresses for wives, stresses from which their husbands are shielded by virtue of the traditional domestic division of labour" (146, p. 58). They support the arguments that women's different role obligations may lead to greater mental health problems and that their work and domestic roles combined may be a likely cause for increased distress. Game and Pringle (320) argue that, in the production environment, work is sexually segregated: light work—which involves less danger, is clean and boring, and involves little movement—is assumed to be appropriate for women. In addition, men usually exercise more control over their work area and tasks.

With many of the symptom groups used in this study, women reported significantly more symptoms than men in the same occupational group. Female supervisors, white-collar (office) workers, and factory workers reported significantly greater emotional symptoms than men in the same groups.

Powles and Salzberg (323) demonstrated correlations in a number of studies between high social status and the most positive health-related behavior, but they issue a caution that the interaction between the sociopsychological and physiological, and effects occurring through health-related behavior associated with the demands of class activities, have not been extensively researched.

The third hypothesis in this study derives from my review of studies (see Chapters 2 to 4) which show that occupational ill-health is likely to occur more frequently toward lower levels in the occupational hierarchy. However, the hypothesis was not completely verified.

Schwalbe and Staples (7) found that a lack of control and job routine were experienced most by blue-collar workers and that these conditions predicted stress and low self-esteem. A significant degree of ill-health resulted. They

concluded that "so long as any labour process denies control and c
whole group of producers, and thereby creates stress, and undermin
then it must produce residual ill-health" (7, p. 598).

In a study of the effects of alienation, stress, and psychological and physical
well-being, Coburn (4) found that men in uninteresting and unchallenging jobs
with little control and variety reported the poorest physical and psychological
well-being. In addition, workers with both underload and overload as alienating
characteristics had lower psychological, physical, and overall well-being than
workers who were not alienated.

Some of the other studies reviewed in Chapters 3 and 4 included Caplan and
coworkers' (83) study which found psychological outcomes to be greatest among
assembly workers. In addition, Frankenhaeuser and Gardell (109) found that
machine-paced workers performing repetitive tasks were more likely to be at risk
for stress and psychosomatic symptoms than workers who performed more varied
tasks and had greater control over the work cycle. The causes for occupational
differences in ill-health have not been researched extensively. Otto (134), Caplan
and associates (83), and others have demonstrated extensive differences in
symptom frequency between those in higher and lower status occupational positions. Labor process writers such as Coburn (4), Garfield (5), Navarro (6), and
Schwalbe and Staples (7) show that management control over the labor process
creates unfavorable and alienating working conditions, especially for lower status
workers.

A high symptom frequency for blue-collar workers can be partially explained
by their work experience. Underutilization on the job itself, with little opportunity
for influence, control, skill utilization, and task variety, is a characteristic of
blue-collar factory work. In addition, management exerts a large amount of
control over the structure of their work processes and this control is accompanied
by a negative climate of attitudes. Little opportunity to exercise control at any
level can lead to responses that create depression and other emotional outcomes
as well as related somatic symptoms. Skilled workers report less negative alienating working conditions and a slightly lower level of symptom awareness.

Managers are the group expected to report the lowest level of symptom awareness, close to the score for skilled workers. In this study they scored more
highly than both supervisors and white-collar workers. Thus, despite reporting a
low level of alienating work, they had a symptom frequency close to the
higher end of the occupational scale. Garfield (5) argues that all occupational
levels experience a certain degree of alienation, managers included. Because
capitalists are subject to competitive relations with labor and with other competing firms and market conditions, they also experience some powerlessness.
Job demand (work overload) is very much higher for managers than for other
occupational groups. Working long hours with significant responsibility for
equipment and people is likely to lead, in the long term, to certain somatic
symptoms.

Supervisors and white-collar (office) workers demonstrate the least symptom awareness. Both groups reported lower levels of alienating work experience than blue-collar workers. These groups have moderate control over their work environment and more opportunities than blue-collar workers to effect changes. Opportunities for task control are relatively high, and the effects of management control of structure and the influence of the climate of attitudes and values are much less evident than for blue-collar workers. Work demand (overload), however, is moderate compared with that of managers.

Employers who spent longer in the one position perceived greater ill-health symptoms. Employees with least mobility are likely to be aware of a greater frequency of symptoms. They generally are from the lowest occupational levels and as a consequence are likely to have least influence over the labor process.

When management control leads to a climate of negative attitudes toward employees—in not treating them with respect, giving them poor feedback, and not appreciating their work—this is likely to lead to a poorer reported state of health. Schatzkin (8) argues that health has meaning to the capitalist as a resource. The existence of negative attitudes toward employees enhances the Marxist view that workers' labor power is a prime commodity and that management's interests in its employees do not extend far beyond gaining that commodity.

Literature by labor process and other writers shows the importance of the effects of task-related alienation (little control over the job and a lack of skill discretion) on ill-health. However, this study found that task did not have an effect on ill-health symptom reporting. Poor work relationships had an effect, but it is only minor. The ill-health effects of a climate of negative attitudes are likely to overshadow the effects of task control and skill utilization and the effects of management control of structure: their likely impact is in creating depression and emotional ill-health.

In Schwalbe and Staple's (7) sample of less than 250 workers from a number of blue- and white-collar occupations, stress was a key determinant of ill-health symptoms. They found, using regression analysis, that stress had a significant effect on ill-health (controlling for self-esteem). Caplan and coworkers (83), Israel and associates (101), and others have also shown the strong effects of stress on ill-health levels. In this study, stress was shown to have a large effect on the awareness of ill-health symptoms. As the level of stress increases, the level of symptom frequency reported is likely to increase. When the demands or resources in the workplace do not match workers' important needs, ill-health is likely to ensue.

The fourth hypothesis in the study is based on literature reviewed in Chapters 3 and 4. A great deal of research has shown stress to be a determinant of ill-health, but much less research has also looked at the effects of low self-esteem and low personal mastery on symptom awareness. The hypothesis that "increased levels of occupational stress will lead to a greater awareness of emotional and somatic symptoms" was verified in this study.

Schwalbe and Staples (7) found self-esteem to have a small but not significant effect on ill-health symptoms. Pearlin and associates (178) compared the effects of self-esteem, self-denigration, and mastery on depression, a stress outcome. They found that sense of mastery had the least effect of all three sociopsychological outcomes of job stress. As argued in Chapter 3, the interaction between stress, low self-esteem, and low level of mastery is quite complex. Israel and coworkers (101) argue that each sociopsychological state could cause, or mediate the effects of, or be an outcome of any of the other sociopsychological variables. As presented in Chapter 6, both low self-esteem and poor sense of mastery are most suitably located together with stress as mediators of the effects of work experience on ill-health.

As Graham argues, evidence shows "there can be little doubt that stress plays an important role in the development of psychological impairment and neurosis (319, p. 154). The role of stress was found to be important in this study for its effects on ill-health. Stress proved to be important for most symptom groups, with the exception of symptoms of analgesics overuse and functional depressive symptoms.

CONCLUSION

Much research into the causes of occupational ill-health has shown that negative work experience can have important health effects, and considerable research has demonstrated that stress is a strong determinant of ill-health symptoms. In addition, there are some positive but lesser effects of low self-esteem and low personal mastery on ill-health. Just how these factors influence each other in the incidence of ill-health, however, is not always clear. Labor process writers argue that management's control over the labor process has led to alienating work environments that lead to stress, other sociopsychological states (such as low self-esteem), and ultimately ill-health. Much of their research, however, focuses on the effects of alienating work at point-of-production activities.

In this study sample, a higher level of education was likely to reduce symptom frequency, and women reported symptoms most often. Only a "lack of consideration," the outcome of a negative climate of attitudes due to management control, had an effect on ill-health. Task-related activities had no effect. According to Littler and Salaman (242) and other labor process writers critical of Braverman's approach, control over the labor process has important effects through other areas of workers' experience apart from control over point-of-production activities. The results of this analysis show that other effects of management's control are important for ill-health at work.

The relationship found between stress and symptom reporting confirms the links, initially established by Selye and supported in a great deal of research, between stress and ill-health. While some studies have shown other socio-psychological variables to have some effect on ill-health, low self-esteem and low mastery were found to have only small effects.

For women, 50% of variations in ill-health frequency are explained by factors in the model. This indicates that those in more disadvantaged class and social positions are likely to have work experiences detrimental to health. In both white-and blue-collar groups, women are likely to experience more symptoms than men. No alienating work factors have an effect on symptom awareness for white-collar workers, and of all sociopsychological factors, only stress has an effect on white-collar employees. Their inability to satisfy certain needs is important in their work experience, and this can be related to identification with middle-class values that tend to regard work as an activity where personal and social rewards can be gained. Working-class socialization and experience are more likely to lead to the view that work is an alienating experience providing little personal reward.

For men, given that the work role is usually their principal life role, it is likely that many important needs require satisfaction at work. If they are not satisfied, ill-health symptoms are likely to result from the stress experienced.

A number of studies identify a lack of skill discretion (skill underutilization and job routine) and a lack of control as the basis of alienating work experience leading to ill-health. In this study, the climate of negative attitudes by management had an overriding effect on symptom awareness. This factor, through demoralizing workers and creating negative sociopsychological consequences, is likely to have prolonged and sustained effects on ill-health. It also represents the capitalist ethos of dominant management with workers experiencing their role principally as a labor resource.

CHAPTER 11

Work, Stress, and Mental Health

This chapter has a number of aims. The first is to determine occupational and gender differences in experience of GHQ symptoms (measured by Goldberg's General Health Questionnaire). The second is to test, based on the causal model of determinants of minor psychiatric impairment, the strength of causal factors in the model. The third is to compare the effects of determinants in the causal model for white-collar and blue-collar employees, and for women and men. This will allow differences in sociostructural location to be tested for in the relationships in the model.

A number of studies have measured the effects of factors at work, including work practices, on mental ill-health outcomes. Long-term effects on well-being can be incurred through a decrease in control by workers (17). A large study undertaken nearly 30 years ago (127) identified a number of negative work factors that contributed to poor mental health at work. Machine pacing has been shown to have adverse psychological effects (83, 110, 116, 328). More recently, studies have found mental health to be worse when there is insufficient decision latitude and discretion for dealing with work demands (66, 105). Depression has also been associated with a lack of control, and is higher for women than men (114).

The influence of other sociopsychological factors on work has been studied. For example, low self-esteem had a stronger effect on measures of depression than did a low sense of personal mastery (178), while poor mastery, together with low levels of self-esteem, had effects on global health measures (101). While a number of studies have found sociopsychological factors to affect somatic ill-health (7, 101), fewer have examined their effects on psychological ill-health.

Relatively few studies have looked at GHQ (psychiatric impairment) levels in the work environment (275–277). The GHQ has been validated for use in occupational settings (279) and is starting to be used more frequently.

The effects of management control over the labor process have been demonstrated in a number of studies which have looked at the outcomes of alienating work characteristics for employees' mental health. Some of these have studied the effects of a lack of control on more general ill-health levels (7). Some have also reported adverse mental health effects resulting from insufficient control (4, 5). However, these writers have focused on the task level, based on the concept established by Braverman (2) through hypotheses on work fragmentation and deskilling. Littler and Salaman (162) proposed a model for dealing with management control at other levels, beyond task; these include the administrative structure and the organizational climate.

Two methods of scoring are used for the GHQ. Likert scales are normally used for a composite score, while the GHQ scoring identifies possible psychiatric cases (294). In occupational settings, however, the identification of cases could be misleading; this method was established primarily for clinical settings: "It is essential to stress that the GHQ recommended here *is* not for the purpose of case identification although it is one of its primary purposes . . . but for comparing levels of psychiatric illness within and between populations" (279, p. 193).

Differences in mental health between women and men have been reported in a number of studies (146, 325, 329). However, relatively few studies have used the GHQ for both women and men at work (275, 279, 330, 331). Differences have been reported between female and male office workers in Hong Kong (275), and lower scores for female workers than for male workers in the United Kingdom (279). However, higher scores for men have also been reported (331). The scale has been used in studies comparing single and married workers (279), and employed and unemployed youth (330–332).

As Broadhead has argued (285, p. 110):

> in the area of psychological health the distinction made between disease and illness is less clear and an indicator of psychological disturbance such as the GHQ score can thus be regarded as measuring something closer to what is evaluated in a clinical situation than is the case with physical morbidity indicators.

Clinical tests have been used extensively in health surveys. A number of studies including the Australian Health Surveys have employed Goldberg's GHQ.

Central to this research was an investigation of structural relationships that contribute to high GHQ scores. This included the effects of negative work factors, stress, and related sociopsychological states for different occupational and gender groups, and the relationship between stress, other related sociopsychological states, and high GHQ levels.

GHQ SCORES, OCCUPATIONAL GROUPS, AND GENDER

The GHQ is an additive 12-item scale with 0 (on a scale of 0–3) the most positive score for each item; the scale score ranges from 0 to 36. An interest in the study was the degree to which managers and lower level workers, particularly unskilled blue-collar workers, showed differences in mental health as represented by GHQ scores. Table 61 gives the range of scores from 11.70 for managers to a high of 14.84 for unskilled workers, supporting large differences between

Table 61

Occupational differences in mean GHQ scores and percentages of cases

Occupational group	Mean (S.D.)	% of cases	n
Managers	11.70 (4.02)	44.6	56
Supervisors	12.29 (3.81)	46.9	49
White-collar (office) workers	12.49 (5.16)	46.7	45
Skilled blue-collar workers	12.28 (4.14)	34.5	29
Unskilled factory workers	14.84 (6.02)	67.3	49
Total	12.73	49.1	228

$F = 3.31; p < .05$

occupational extremes. Supervisors, white-collar (office), and skilled blue-collar workers all reported scores close to the sample average and closer to managers' scores than to factory workers' scores. Differences between groups are significant ($F = 3.31; p < .05$). The greatest differences are between unskilled blue-collar workers and three other occupational groups (managers, supervisors, and white-collar (office) workers). These are significant on a multiple range test using the SNK procedure.

Tables 61 and 62 present case identification for the manufacturing organization sample in the study. GHQ scoring uses 0, 0, 1, 1 for items (294) and a cutoff score of 1/2 to determine possible psychiatric cases. This was a useful measure of the range of cases across occupational groups, but less useful as an interpretation because case identification is most appropriate in clinical settings. Large numbers of cases were identified in the unskilled factory group, but a relatively large number of cases predominated through the whole sample. Cases range from 34.5% for skilled blue-collar workers to 67.3% for unskilled factory workers (Table 61). These rates are high for a work environment. Managers, supervisors, and white-collar (office) workers all have similar case levels.

Differences between women and men in mental health at work are of particular interest. Table 63 shows that women reported significantly higher scores (13.94) than men (11.83) ($F = 11.11; p < .01$). For women, managers have the lowest score (10.89) and unskilled factory workers the highest (15.84). However, the differences are not significant ($F = 1.82$). For men, white-collar (office) workers have the lowest score (10.57) and unskilled factory workers the highest (12.94) ($F = .74$; not significant). The range of scores for women is more than twice that of men. Surprisingly, the largest (but not significant) difference between women's and men's scores is for white-collar (office) workers ($F = 2.93$). As Table 62 shows, the range of cases for women is from 44.4% for managers to 71.9% for unskilled blue-collar workers. This is higher than the range for men, with 32% for

Table 62

Percentages of cases for women and men within occupational groups

	Women	Men
Managers	44.4	44.7
Supervisors	57.1	39.3
White-collar (office)	45.2	50.0
Skilled	50.0	32.0
Factory	71.9	58.9
Total	56.4	43.5

Table 63

GHQ scores for women and men within occupational groups

	Women		Men		
	Mean (S.D.)	n	Mean (S.D.)	n	F
Managers	10.89 (4.78)	9	11.85 (3.90)	47	0.43
Supervisors	13.05 (3.88)	21	11.71 (3.73)	28	1.48
White-collar (office)	13.35 (8.76)	31	10.57 (2.79)	14	2.93
Skilled	14.75 (6.13)	4	11.88 (3.76)	25	1.70
Factory	15.84 (6.49)	32	12.94 (4.63)	17	2.67
Total	13.94 (5.71)	97	11.83 (3.83)	131	11.11*

$*p < .01$

skilled blue-collar workers as the lowest score and 58.9% for factory workers as the highest.

DETERMINANTS OF GHQ SCORES

The study aimed to identify which of a number of individual and work characteristics were the strongest determinants of mental ill-health as measured by the GHQ. A regression equation containing four blocks of variables gave regression coefficients that are the total effects of variables resulting from a causal modeling approach (Table 64). Standardized and unstandardized coefficients are displayed, but only the standardized regression coefficients are discussed in the analysis. In

Table 64

Determinants of GHQ scores: regression coefficients

Independent variables	Dependent variable, GHQ score	
	b	beta
Ascribed and achieved statuses		
Number of children	0.00	−.01
Primary education	3.36	.04
Apprenticeship	−5.70	−.15
Tertiary education	−2.67	−.09
Partner not working	0.32	.01
No partner	1.53	.05
Age	0.58	.04
Gender	−4.02	−.15
Adjusted *R*-squared		.03
Occupational status		
Managers	0.57	.02
White-collar (office) workers)	−0.89	−.03
Skilled workers	2.19	.05
Factory workers	5.02	.15
Years in the one position	1.15	.10
Adjusted *R*-squared		.04
Negative work factors		
Role conflict	1.06	.06
Poor relationships	2.50	.13
Lack of freedom to control	−0.30	−.02
No skill discretion	2.01	.10
Overwork	−0.29	−.01
Lack of consideration by management	5.47	.28*
Adjusted *R*-squared		.12
Sociopsychological responses		
Low self-esteem	7.50	.26*
Stress	10.57	.38*
Lack of mastery	4.07	.14
Adjusted *R*-squared		.38

*Significant at < .05

a multivariate ANOVA test on intercorrelated independent variables, no significant interaction effects were found.

There are no real effects of ascribed or achieved statuses, or of differences in occupational level. In fact, only one negative work factor, "lack of consideration by management," has a moderate effect (beta = .28). Of the sociopsychological responses, stress (.38) has the largest effect of any of the variables in the model. The experience of stress at work is the most effective predictor of the variables included in the model for differences in GHQ levels. Low self-esteem (.26) has a lesser effect, but shows that differences in self-esteem are effective determinants of GHQ scores. Overall, the R-squared (.38) indicates that many of the influences that affect variations in GHQ scores are factors not included in the model, highlighting the need for further exploratory research into work factors that may influence mental health as measured by the GHQ.

Separate analyses were conducted for white-collar employees (including managers, supervisors, and white-collar (office) workers) and blue-collar employees (skilled and unskilled factory workers) (Table 65). All variables not having a moderate to large effect are omitted, with the exception of all negative work factors and sociopsychological variables. The aim was to look for separate patterns of factors affecting GHQ scores for each of these groups. For blue-collar workers, women were likely to have moderately higher levels of psychiatric impairment than men (beta = −.35 and significant). Changes in stress scores were the largest determinant of GHQ levels for blue-collar workers (.40), demonstrating the strong effects of high stress on GHQ scores. Level of stress at work could have important effects on mental health for this group. Self-esteem levels had a moderate (.32) effect on GHQ levels. The level of stress on the job, self-esteem, and (to a lesser extent) a lack of consideration by management were all important for the mental health of blue-collar employees. Almost half of the variations in GHQ scores are accounted for by the model (R-squared = .47).

The pattern of determinants for white-collar employees centers on sociopsychological factors alone. Self-esteem levels had the greatest influence on mental health levels (.34), while stress levels (.27) had a lesser but important effect. Variables in the model have little influence in explaining variations in the white-collar workers' GHQ scores (R-squared .24). A lack of personal mastery has no important effects for either blue- or white-collar workers. The lack of effect of negative work factors on psychiatric impairment for white-collar workers is surprising, and indicates that no one specific work practice measured is likely to have deleterious effects. Research is needed into determining what other factors are likely to account for variations in scores.

In order to assess the effects of gender on psychiatric impairment, separate analyses were performed for women and men (Table 66). For women, poor relationships at work" (beta = .30) is the only factor in the model to have a moderate to large (and significant) effect on GHQ scores. Surprisingly, none of the sociopsychological outcomes (such as stress and self-esteem) had a

Table 65

Determinants of GHQ scores by occupational group

Independent variables	Dependent variable, GHQ score (beta)	
	White-collar	Blue-collar
Gender	−.10	−.35*
Negative work factors		
Overwork	−.05	.09
Lack of freedom to control	−.01	.06
No skill discretion	.00	.14
Poor relationships	.17	.08
Lack of consideration by management	.17	.32
Role conflict	−.01	−.03
Sociopsychological responses		
Stress	.27*	.40*
Low self-esteem	.34*	.32*
Lack of mastery	.10	.15
Adjusted *R*-squared	**.24**	**.47**

*Significant at < .05

sufficiently large effect. It is of interest that the quality of personal relationships should be the major factor affecting women's GHQ level. A number of factors have a reasonably strong effect on GHQ scores for men. A lack of consideration by management (.34) is the only negative work factor. For men, stress at work has a large effect (.47): changes in stress scores were likely to be important in explaining variations in GHQ scores. Self-esteem scores (.36) had a lesser but important effect on levels of psychiatric impairment. As with women, about one-third of men's GHQ scores are explained by factors outside the model (*R*-squared = .32 and .36, respectively).

GHQ: THE EFFECTS OF WORK EXPERIENCE, SOCIOPSYCHOLOGICAL OUTCOMES, AND JOB SATISFACTION

A particular interest in the study was the relationship between a number of variables and GHQ levels of employees. Low, moderate, and high levels of negative work, stress, low self-esteem, lack of personal mastery, and job satisfaction were calculated based on a statistical breakdown of the scales (Table 67). Differences in the levels of negative work experienced are associated with large

Table 66

Determinants of GHQ scores by gender

Independent variables	Dependent variable, GHQ score (beta)	
	Women	Men
Negative work factors		
Role conflict	.15	.03
Poor relationships	.30*	.06
Lack of freedom to control	−.06	.03
No skill discretion	.10	−.07
Lack of consideration by management	.21	.34*
Overwork	−.12	−.02
Sociopsychological responses		
Stress	.24	.47*
Low self-esteem	.21	.36*
Lack of mastery	.26	−.02
Adjusted *R*-squared	**.32**	**.36**

*Significant at < .05

differences in psychiatric health. Those with low levels of negative work experience reported a GHQ level of 10.52: the high negative work group recorded a
score of 19.12. GHQ differences between levels of negative work experience are
large and significant ($F = 23.80, p < .01$).

The largest differences occur between levels of stress, with GHQ levels ranging from 10.52 for the low stress category to 18.74 for the high stress category
($F = 32.04; p < .01$). Differences in self-esteem (sense of self-worth) have less
dramatic effects on scores for psychiatric impairment ($F = 30.95, p < .01$).
No employees fell in the lowest esteem (high levels of low esteem) category.
Employees in the high esteem (low levels of low esteem) category recorded a
mean GHQ score of 11.66, while those with moderate esteem scored a level of
15.41. Differences between low, moderate, and high levels of lack of mastery
were large ($F = 24.11; p < .01$). Employees experiencing a high degree of
personal control (low levels of lack of mastery) over their environment had
considerably lower GHQ scores (10.77) than those experiencing low personal
control (high levels of lack of mastery) (19.55). Differences between scores for
high, moderate, and low levels of negative work, stress, and lack of mastery all
were significant on multiple range tests using the SNK test.

One measure of particular interest in relation to mental health at work not
included in the previous analyses was job satisfaction. Scores for job satisfaction

Table 67

GHQ scores for negative work, sociopsychological states, and job satisfaction

	Mean (S.D.)	n
Negative work		
Low	10.52 (3.35)	56
Moderate	12.87 (4.65)	126
High	19.12 (6.52)	17
Total		199
$F = 23.80*$		
Stress		
Low	10.52 (3.35)	56
Moderate	13.95 (4.48)	75
High	18.74 (7.06)	19
Total		193
$F = 32.04*$		
Low self-esteem		
Low	11.66 (4.41)	170
Moderate	15.41 (4.89)	63
High	— (—)	—
Total		223
$F = 30.95*$		
Low mastery		
Low	10.77 (3.76)	92
Moderate	13.49 (4.68)	132
High	19.55 (6.11)	11
Total		235
$F = 21.59*$		
Job satisfaction		
Satisfied	11.94 (4.17)	189
Dissatisfied	16.14 (6.17)	28
Total		217
$F = 21.59*$		

$*p < .01$

and job dissatisfaction were derived using a simple 12-point scale measuring six points of satisfaction and six of dissatisfaction. Only a small number of employees reported job dissatisfaction, but their GHQ scores on average were four points higher than those of satisfied employees. Differences in GHQ scores between both groups of employees were moderate ($F = 21.59$; $p < .01$). Negative work, high levels of stress, low self-esteem, low personal mastery, and dissatisfaction with work are all characterized by high GHQ levels.

DISCUSSION

Numerous studies have focused on mental ill-health consequences for employees at work. Some studies have reported incidences of depression (83, 100, 114, 178) and anxiety (83, 98, 333). A relatively small number have reported anxiety, depression, and related measures for different occupational groups (83, 127). Others, however, have also linked personality characteristics to mental health outcomes at work (334). Clearly, what is needed is a closer evaluation of the types of working experiences that can contribute to mental ill-health, identified for a range of occupational groups.

Two particular features need further investigation. First, a number of studies have shown mental ill-health to result from specific aspects of work organization and practice, but further research is needed to determine more clearly how different aspects of work organization contribute to psychiatric impairment. The effects of opportunities for control, of role conflict, underutilization of skills, responsibility, organizational climate, role loss, fragmented work, and the effects of social environment (65, 66, 149, 335–340) have all been reported. Few studies, however, have looked at the specific effects of work factors on GHQ levels. The second feature is differences in mental health outcomes for different occupational groups, as yet reported by relatively few studies (133, 134, 336). This study has attempted to address these issues, as well as the role of stress and other sociopsychological states) as moderating variables for minor psychiatric impairment.

Some occupational differences were found in this study. Large differences are evident between GHQ scores of managers and those of employees in lower occupational groups. Symptoms of mild psychiatric impairment are experienced most by unskilled workers. One of the few studies of occupational differences has shown that stress, as measured through GHQ scores, was higher for basic-grade prison officers than for senior officers (insufficient details on the method of scoring (0–48) were given for an adequate comparison (341).

The GHQ scores in this sample were compared with high scores in other studies. For example, Graetz (331) found an average score of 11.48 for unemployed groups in a study of employed and unemployed 16–25 year olds; this was comparable to blue-collar scores in this study. Graetz's study showed an average GHQ score three points lower than that in the current study sample. Lower scores were also found in three studies (means for women 8.53, and for men 8.80) than in my study. Even lower scores have been reported for production workers in the oil industry (342): a sample of onshore production employees had scores of 7.64, while offshore workers had scores of 8.75. In a sample of nurses, comprising mainly women, those using low engagement and low disengagement as coping approaches had an average GHQ of 10.45, whereas those using high engagement and low disengagement had the highest average score, 14.55 (343). The high scores were comparable with those of the blue-collar workers in my study sample. Slightly lower GHQ scores were reported for female and male office workers in Hong Kong than for white-collar (office) workers in my study sample (women 24.92 and men 22.50, using a 1, 2, 3, 4 scoring method) (275). High scores in the study reported here may be due to onerous demands in the manufacturing work environment.

Women's GHQ scores were higher than men's in the study. Other authors have reported similar differences. Lam and colleagues (275) found a score of 24.92 for female white-collar (office) workers and 22.50 for male office workers. Graetz (278) reports a GHQ score of 10.67 for women, significantly higher than for men (9.38), in a study of 16- to 24-year-old Australians. However, lower scores were reported for women (8.53) than men (8.80) in a British sample (279), and in an unemployed sample from a closed wood-processing factory in Finland (2.8 and 3.7, respectively; scores based on a scale range of 0–12) (344). However, women's scores were higher than men's in a control group of employees from another Finnish wood-processing factory (2.3 and 1.1, respectively), as reported by Viinamaki and coworkers (344).

A major interest in the study was to determine which factors, particularly occupational level and work organization and practice, contribute to differences in levels of minor psychiatric impairment. A number of authors have found that control over work and participation in decisions have important effects on mental health (65, 66, 108, 127). Much research has conceptualized control as influence over the task or job, in order to better understand mental ill-health at work. Numerous studies have measured control under machine-pacing conditions, for example, and its effects on mental health. Labor process researchers (4, 5, 7) have employed measures of control over task activities to explain the effects of management control over work activities on a number of levels of health, not only mental health. One study, however, showed that measures of alienation (such as repetition and lack of autonomy) correlated "at best only weakly with the indices of general well-being" (4, p. 46). Measures of "subjective alienation" (interest and challenge in the job), however, correlated highly with well-being. Little research has tested the effects of other types of management control (that is, lack of control by employees other than control over the task). A lack of consideration by management, emphasizing not listening to employees, giving no feedback, and showing no appreciation, had important effects on GHQ scores in my study.

As numerous reports have shown, stress mediates the effects of negative work on a range of health outcomes (see Israel and coworkers (101), and Schwalbe and

Staples (7)). Few studies however, have demonstrated how stress mediates the effects of work on mental health outcomes, especially minor psychiatric impairment. In this study I have shown that stress is an important experience in explaining GHQ scores—but only for the men in the sample. In fact, a number of other variables, including a lack of consideration by management and low self-esteem, helped to explain men's GHQ scores. Men's roles and identities tend to center around work. As a consequence, their mental health is likely to be integrally tied to their work role, with impairment resulting from work-related pressures.

Women's mild psychiatric impairment was found to be related to the quality of work relationships. A number of writers have drawn attention to the effects of multiple important roles for women (see 325), including employee, homemaker, and family caretaker, and in some cases work can provide opportunities to create relationships away from the demands of these other roles.

Other studies have looked at the roles of sociopsychological states such as self-esteem and personal mastery in contributing to depression, among a number of outcomes (e.g., 178). I found self-esteem to have important effects for men only.

The relationship between dissatisfying work and high GHQ scores found in this study has been confirmed elsewhere. In a study of employed and unemployed school leavers, dissatisfaction with employment was reported as a mental health risk (dissatisfied employees reported higher GHQ scores than the unemployed in the sample) (331). Similar differences were previously found with another Australian sample (330). "The psychological benefits in employment are not delineated across the board, but are confined to those who are satisfied with their work" (331, p. 721).

CONCLUSION

Of course, many of the causes of differences in GHQ scores between occupational groups may be found in extraorganizational factors, such as lifestyles and patterns of socializing. The low R-squared values for some of the groups suggest a need for more extensive research into these groups. The study has identified some important differences in the causes of psychiatric impairment between women and men, and between blue- and white-collar workers. It has also shown few negative work factors to have effects on psychiatric impairment. However, the work stress experienced by employees is very important in understanding the occurrence of mental ill-health at work.

Conclusion

This book has looked at the effects of management's control over the labor process in capitalist society. It has examined the differences in occupational stress between blue-collar and white-collar workers and between women and men, and the effects of management's control on alienating work.

As workers fail to satisfy important needs at work they are likely to experience stress, and an alienating working environment is likely to produce low levels of self-esteem and personal mastery. These negative sociopsychological states in turn are likely to lead to physical and psychological ill-health. Management's control is most likely to result in blue-collar workers having the least control and influence and the highest levels of stress and ill-health.

This study of a large manufacturing company set out to test the influence of sociostructural factors on occupational stress and ill-health. Based on a review of the literature in Chapters 1 to 5, I developed a causal model of the major contributing factors. The main sociological problem addressed was that working-class (blue-collar) groups are likely to experience disadvantageous stress and ill-health at work. This is largely due to their more alienating work experience, deriving from their having less ability to exert control over the labor process. In addition, they are likely to have fewer sociopsychological resources to deal effectively with negative work experience.

There were several outcomes of the study, including the results of tests of the four major hypotheses and their implications for stress and ill-health research. In this chapter I summarize the characteristics of alienating work, stress, personal resources, and ill-health for the different occupational and gender groups, the implications for a further understanding of important elements in the stress process, and the implications for stress research of accounting for differential class relations and class conflicts, and specifically for using a labor process approach. I also address some of the implications for further research and some policy implications for addressing occupational ill-health in industry.

MAJOR EXPECTATIONS

Four hypotheses were tested: first, that lower occupational (blue-collar) groups will experience greater stress than higher (white-collar and management) groups;

241

second, that lower occupational groups will experience the lowest levels of self-esteem; third, that as occupational status decreases, symptom frequency will increase; finally, that the experience of stress will be an important determinant of ill-health symptom frequency. The higher the level of stress that workers experience, the more likely they are to report a higher ill-health symptom frequency. I will discuss the implications for a labor process understanding of stress and ill-health.

The first major hypothesis, "as occupational position moves toward lower (blue-collar) levels, stress levels will increase," was not completely supported. There were several reservations. Managers, as expected, had one of the lowest levels of stress. Factory workers, the lowest occupational group, had the highest stress levels. Caplan and coworkers (83), Otto (134), Schwalbe and Staples (7), and Kohn and coworkers (133), and others have shown that unskilled blue-collar workers experience greatest stress. Schwalbe and Staples argue that class location is an important link in the experience of stress. Those in working-class positions have only their labor power to sell; they have little opportunity for influence and control and fewer social and personal resources with which to deal with inordinate demands at work. The work conditions of lower occupational groups in the sample were characterized by the most alienating work. White-collar workers and skilled blue-collar workers were more likely to report lower stress levels than managers. Skilled workers were likely to have a stronger base of personal and craft-related resources for dealing with alienating work demands. Managers were likely to experience slightly higher stress. However, working in an unskilled factory job such as a machine tender or floor help was likely to result in the highest levels of stress of all occupational groups.

The second major hypothesis, "as occupation level decreases, levels of self-esteem will decrease," was supported. Managers were likely to have the most positive sense of self-esteem. Factory workers, however, experienced a moderate to large degree of lower self-esteem. Factory work is conducive to low positive appraisals, low self-reflected appraisals, and poor sense of competence. Schwalbe and Staples (7) argue that class positions which lead to a lack of control and job routine at work are likely to result in a negative sense of self-worth. This was most evident in factory work in the study sample. The experience of low self-esteem is closely related to class and occupational membership.

The third hypothesis was that "as occupational status decreases, symptom awareness will increase." Factory workers recorded the highest frequency of somatic and emotional symptoms. Managers reported the lowest. Differences in reported symptom frequency between occupational groups were moderate, largely supporting the hypothesis. Labor process writers such as Schatzkin (8), Navarro (6), and Garfield (5) argue that employees' health is affected by different class experiences at work. My study of a manufacturing company shows that in most cases workers reporting the greatest alienating working conditions also reported the poorest levels of health. Management control over the labor process places the greatest restrictions on blue-collar (especially factory) workers' opportunities to influence changes and to use discretion at work. These groups reported the greatest ill-health.

The final major hypothesis was that "increased levels of occupational stress will lead to a greater awareness of emotional and somatic symptoms." Selye, a major contributor to stress theory and research, was one of the first to establish a definitive link between stress and ill-health. A large amount of subsequent research has substantiated this link. Stress was a major determinant of somatic and emotional symptom frequency in the study sample. The greater the degree to which employees felt troubled, worried, or bothered about work, the more likely symptom frequency would increase. Coburn (4), Garfield (5), and Schwalbe and Staples (7) argue that, with increased alienation, working-class employees meet fewer important needs at work, are likely to feel greater stress, and as a consequence have poorer health. Caplan and coworkers (83), Israel and coworkers (101), and House (307) are some of the many other authors to demonstrate causal relationships between stress and ill-health.

This study controlled for the effects of other sociopsychological states (self-esteem and mastery) on ill-health symptoms. Stress was by far the most important of all sociopsychological determinants, and the strongest predictor out of all the status variables and negative work factors.

OCCUPATIONAL AND GENDER DIFFERENCES IN ALIENATING WORK, STRESS, PERSONAL RESOURCES, AND ILL-HEALTH

The experience of alienating work, stress, personal sociopsychological factors, and symptom frequency is presented for each occupational group. White-collar groups (managers, supervisors, and white collar (office)) and blue-collar groups (skilled and factory workers) are discussed separately. White-collar groups have been considered as a composite group for some analyses, for two major reasons. First, as literature reviewed in the earlier chapters argued, white-collar (office) workers and all levels of managers tend to be more supportive of management's goals than do blue-collar workers. They also tend to have more values in common compared with blue-collar workers, many of whom have a history long steeped in the labor movement. Second, there were insufficient cases in individual occupational groups to perform a meaningful analysis through regression analysis. White-collar and blue-collar groups therefore have been considered as separate, composite groups.

White-Collar Groups

Managers generally experience the least alienating working conditions and lower levels of stress than most other groups. Personal resources mitigate the effects of negative work experiences and interact with stress to mediate its effects on outcomes of ill-health. In the study sample, their self-esteem levels were highest, as were their levels of personal mastery, giving them advantages in dealing with the effects of negative work. However, working as a manager was likely to be a moderate determinant of stress and ill-health compared with other groups.

Due to their large degree of control over the labor process, managers experience very real advantages as a result of their work, including both personal resources and somatic and psychological health. This is not to say that managers do not experience some deleterious effects of work in capitalist society. Moss (208) argues that managers work under considerable pressure. Some experience a degree of powerlessness due to the demands of a board of management and shareholders. In addition, argues Moss, managers can be under pressure relating to their role performance and personal expectations, which are likely to be highly demanding. In the organization in the study, managers were under pressure to meet the expectations of an overseas board of directors and management. However, in comparison to other occupational groups, they had stress, esteem, and sense of personal control levels that were quite positive, as well as quite beneficial health levels.

Supervisors are included in white-collar groups as lower level management employees. Hill (244) argues that those in lower level management tend to be supportive of management prerogative; they also are likely to support many of the values of more senior management, although this is not always the case. Supervisors in the sample experienced a moderate degree of alienating work, between the two extremes of managers and factory workers. However, they were likely to experience higher levels of stress than all other groups except factory workers. Working in a supervisory position is a moderate determinant of stress, low self-esteem, and mastery compared with other occupational groups. However, supervisors were most likely to report the lowest levels of ill-health symptom frequency and of psychiatric impairment, except for white-collar (office) workers.

White-collar (office) workers, according to Hill (244), are likely to support management goals and values to a greater degree than blue-collar workers. They enjoy good physical working conditions and opportunities for mobility. White-collar (office) workers experienced a lower level of work alienation than other groups in the study sample. White-collar (office) employees had work with a low alienating content and were likely to have high levels of self-esteem and moderate levels of personal mastery, compared with other occupational groups. They were likely to experience conditions that lead to the least stress. Opportunities to fulfill important needs are greatest for this group of workers. Their ill-health symptom awareness and level of psychiatric impairment were the lowest of all occupational groups. Middle-class lifestyles and values are likely to ensure positive health-related behaviors.

Blue-Collar Groups

Blue-collar workers (semi-skilled, skilled, and unskilled) have been treated as a composite group, compared with white-collar groups, for some analyses. They are likely to form a separate class group from white-collar employees, being more likely to share interests and value opposing those of management. Skilled and semi-skilled employees are members of trade groups within the organization. They have affiliations with specific skills-based trade unions and in many cases have apprenticeships.

Skilled employees experienced a degree of alienating work second only to that of unskilled factory workers. However, they were likely to experience greater stress than only white-collar (office) workers. A lack of personal resources (low self-esteem and low levels of mastery), however, was likely to be high. Only factory workers were likely to be more lacking in these resources. Skilled workers were also most likely to report a slightly smaller frequency of ill-health symptoms than factory workers, the group reporting poorest health, and were likely to report the second greatest level of psychiatric impairment. Factory workers reported greatest impairment.

The unskilled factory workers in the sample occupied machine tender positions and floor help jobs. They reported by far the greatest alienating working environment and conditions, and were most likely to experience the highest levels of stress. Their work allowed them fewest opportunities to meet important needs. Self-esteem and personal mastery were likely to be lowest of all the occupations, and ill-health frequency awareness the greatest. Their levels of psychiatric impairment were also likely to be the greatest.

Gender Groups

Game and Pringle (302), Baxter and colleagues (306), and others argue that women experience considerable disadvantage at work. They normally occupy lower level jobs than men, and as a result earn less and achieve less mobility due to fewer promotion and career advancement opportunities. Many of the jobs occupied by women have already been deskilled and vacated by men. Other jobs (such as secretarial and typing) are traditionally women's jobs.

The only occupational group in the sample in which women reported more alienating work than men was factory workers. Factory women reported more stress than factory men, but the difference is not significant. This lack of gender difference in stress is confirmed in a number of studies. However women's personal resources of self-esteem and mastery were more likely to be lower than men's when compared within occupational groups. But it is in reporting of ill-health symptom frequency that we see great disadvantages for women. These poorer levels of physical and emotional health for women have been reported in many studies. Female supervisors, white-collar (office) workers, and factory workers reported significantly more symptoms than men in the same occupational groups. Several explanations of ill-health outcomes for women are possible, some of which account for women's important domestic role requirements as well as gender differences in socialization and health requirements for particular role performance. Some female white-collar (office) workers and factory workers report a greater level of psychiatric impairment than men from the same occupational groups, but these differences are not significant.

THE RELATIONSHIP BETWEEN STRESS, PERSONAL RESOURCES, AND ILL-HEALTH

The model of stress and ill-health used in this study looked at the inability of demands and resources at work to meet important needs of workers. The model also examined the effects of work demands and resources on the efficiency of personal sociopsychological resources (low self-esteem and low sense of mastery). Negative emotional states (stress) and the robustness of personal resources (low self-esteem and low mastery) were tested in turn for their effects on the frequency of ill-health symptoms and on levels of psychiatric impairment (GHQ levels).

Future stress research should incorporate a number of factors external to the work environment, such as the effects of other life roles, lifestyles, patterns of health behavior, and access to important life resources. These were beyond the scope of the current study, but in future research could provide valuable interactive effects for the influence of work conditions.

Literature based on a labor process approach shows blue-collar workers are likely to experience the greatest stress and ill-health disadvantage at work. This study, based on a labor process approach in explaining the occurrence of stress and ill-health, has shown that working-class (blue-collar workers) had the greatest health disadvantage. Management exerted greatest control over these workers, and they had the greatest degree of alienating work, highest stress, and greatest health disadvantages. The stress process model shows that alienating work contributes to high stress and diminishes the self-concept through lowering self-esteem and personal mastery. Negative work experience, together with the resulting stress and (to a lesser extent) erosion in the self concept, led to poor somatic, emotional, and mental health. These results, using the model of alienating work–stress–ill-health, have been confirmed in various ways in a number of studies.

There were large differences in stress outcomes between factory and white-collar (office) workers in the sample. Surprisingly, differences between factory workers and managers for stress outcomes were small to moderate. This is possibly due to the large workloads and responsibilities of managers in the sample. Differences in self-concept outcomes, however, were moderate to large for these two groups. Factory work was predictive of the poorest health and lowest psychiatric health levels, but there were small differences between factory and managerial jobs as health determinants. White-collar (office) workers were likely to have the lowest symptom frequency and least psychiatric impairment. Labor process studies and other research have shown these class differences in stress, and in somatic, emotional, and mental ill-health. Some similar results emerged in class differences in this study. For both white-collar and blue-collar groups, some alienating work factors were important predictors of stress. Alienating work, however, was not a strong determinant of ill-health symptom awareness for either group.

Stress was an important determinant only for white-collar workers' symptom awareness; low self-esteem and poor mastery had no real effect on symptom levels for either group. Meeting needs had more important health consequences for white-collar employees. Blue-collar workers may have lower expectations about the needs they can satisfy at work, based on past work experience and earlier class socialization.

THE LABOR PROCESS, BRAVERMAN, AND STRESS AND ILL-HEALTH: THE IMPLICATIONS OF CLASS DIFFERENCES

According to Willis (241), Braverman's work has three omissions. First, Braverman's account is a purposely "one-sided account of class formation" (241, p. 11). While he is aware of workers' abilities to react to management control, it looks as if management achieves its end. Second, Braverman ignores the role of ideology as important to an understanding of capitalism. Third, the role of the state is underdeveloped in Braverman's work. His focus on the type of changes in labor process organization is limited. Despite these criticisms, Braverman has provoked some important discourse on the class conflicts underlying the labor process. While his theory has important implications for an understanding of the organization of the labor process, I argue that there are certain limitations in confining the extent of the effects of management control over work to point-of-production activities. As Child (161) suggests, the process of management policy formation and decision-making involves organizing structure in order to ensure that managerial objectives are accomplished. Further, as Littler and Salaman (162) note, management influence extends to the employment economy. This implies that most aspects of work that result from management's attempts to meet its objectives (its objectives often being at odds with the needs and objectives of workers) result in management's gaining further control over the labor process. This includes the structure of work relations and the climate of attitudes expressed by management toward workers. Many other aspects of work—including compensation schemes, promotion and career opportunities, and physical conditions of work—also result from management attempts to maximize capitalist objectives. These have not been included in this study.

From the point of view of occupational stress and ill-health, the whole range of negative or alienating working experiences can be regarded as important causes of stress and ill-health, emanating from management control over the labor process. Many studies have focused on the effects of negative task conditions on stress and occupational ill-health. Many labor process authors have conceptualized alienating work as alienation on the job itself (a perceived lack of control, job routine, and work underdemand). Such studies have found, in the main, that these are predictors of stress and of ill-health. However, few studies have treated management control of the labor process as having broader effects on work experience.

In this study of a manufacturing environment, alienating conditions on the job itself (a lack of control and lack of skill utilization and variety) have small effects on stress relative to the effects of structural control by management (role conflict and poor relationships), and especially the effects of a negative climate of attitudes by management. The lesser effect of task-related factors on stress (and ill-health) is also apparent for blue-collar workers as a group, as well as for white-collar groups. This does not imply that these factors are not important sources of stress and ill-health. However, their effects have been overshadowed by the daily effects of management's bureaucratic work organization (structure) and negative attitudes toward employees. These factors probably represent the most irksome features that workers carry away from the workplace. For workers, being treated as not making a worthwhile or valued contribution, seldom receiving thanks or appreciation from management for work effort, and receiving little respect or consideration from management represent an ideology of management dominance. Negative management attitudes toward workers had the greatest effect of all negative work experience on stress, low esteem and mastery, and ill-health symptom awareness.

The relatively small effect on stress of a lack of control and opportunity to use skill discretion, when included together with other measures of management control, can also be explained by workers needing to gain some personal sense of worth and value out of work that is inherently meaningless. Shostack (195) suggests that blue-collar workers need "self-esteem protection." He argues that over many decades blue-collar workers have experienced great dissatisfactions and have learned to ignore their effects or even to deny their existence. Stressors that seem to be beyond the control of workers and for which there seems to be little hope of change may eventually be regarded as affecting needs that are not important. A lack of control, skill underutilization, and routine work are characteristic conditions of blue-collar work. However, negative attitudes, expressing management's dominance of workers, promote a level of self-worth that is continually reinforced as low. The reality of being a labor power commodity is evident directly on the shop floor through negative responses from management.

Karasek (65, 66) describes two types of stress studies: those that examine the effects of a lack of control over work and those that deal with the effects of other

work stressors. My study indicates the need to account for a range of alienating work experiences that result from management control of the labor process in order to more fully understand the effects of such control on workers' stress and ill-health. I have found that occupational position and alienating work experience variables have important predictive value for stress and personal resources as well as for somatic and emotional symptom frequency. This supports important findings from much labor process research.

The capitalist manager is most interested in achieving capitalist objectives; employee health is an expendable commodity in this process. According to Schatzkin (8), those with fewest resources and least power are most likely to experience harmful effects on health from their work. In order to maintain productivity, a certain amount of health is required: "the Capitalist is simply not interested in the level of health beyond this" (8, p. 215).

IMPLICATIONS OF THE STUDY

In this study of a manufacturing organization, the experience of stress was greatest for blue-collar unskilled workers, and they also reported the greatest levels of ill-health. While managers reported low levels of stress and a low symptom awareness, white-collar (office) workers were likely to report least stress and most positive somatic and emotional health levels. This research supports the findings of a number of labor process studies, that alienating work resulting from management control is likely to lead to increased stress and consequent ill-health.

Current research into occupational stress has followed two fairly predictable paths. Research based on a physiological orientation has examined personality differences in order to explain differences in stress responses. It has followed a more management orientation to the causes of stress and ill-health at work, ranging from blaming workers for becoming ill to identifying poor work conditions and poor job design as the principal causes. These studies tend to be conducted outside a comprehensive sociostructural and political framework. Sociological studies, on the other hand, have generally concentrated on providing a structural explanation for the experience of stress and are more likely to seek to locate the causes of stress and ill-health in the nature of workplace relations. Labor process writers have more recently developed a more comprehensive approach: they have identified management's attempts to exert control over the labor process as the main cause of stress and ill-health.

This study identifies certain deficiencies in approaches taken in current research and supports the orientations of the more recent structurally based labor process approaches. We need to look for the causes of occupational stress beyond conditions in the workplace: in the nature of labor and management relations and in the broader role of the state in supporting and maintaining various forms of management control. Management control is a condition that precipitates the experience of stress and ill-health at work, and thereby further focuses our attention on the concept of control as important in understanding stress. A relatively small group of studies have examined the effects of control on stress and ill-health, but many of these, with the exception of mainly labor process studies, have failed to operationalize control as part of the broader sociopolitical context within which work is carried out.

The aim of this book has been to go further, to operationalize the concept of control at a number of important levels at work: control of the task, bureaucratic or structural control, and management's negative attitudes toward employees that derive from its control. Labor process studies operationalize control as alienating work at the level of the task only, thus ignoring some of the more alienating outcomes of management's control under capitalism. The majority of other studies that measure the effects of work factors beyond the level of task fail to conceptualize them as part of a wider set of work relations and work structures.

Stress needs to be seen as part of broader sociopolitical processes, and a Marxist framework should be extended to incorporate a more total picture of workers' experiences in the workplace. As one of the few studies to emphasize occupational and gender differences, the study reported in this book points to our limitations in understanding the causes and consequences of occupational and more general class differences in the experience of stress and ill-health.

There is now a greater emphasis on workplaces being highly productive in order to survive, which places a premium on competitiveness. In times of low unemployment it was relatively important for management to attempt to satisfy workers' needs in order to reduce labor turnover. This is less important now. Occupational stress arises when few important needs are met at work or when resources at work are insufficient to meet the demands made on employees. With high competitiveness and high levels of unemployment, the demands on management to satisfy the needs of groups such as unskilled blue-collar workers are not high.

In Australia, occupational health and safety legislation has ensured that certain minimum levels of health and safety are maintained at work. However, with government budgetary constraints, these initiatives are constantly under threat of being minimized. And with the concentration of conservative state governments, there are policies which at times undermine some basic rights to compensation.

One feature of recent economic development has been a growing ideology of both government and managements that *all* employees are best served by working toward corporate goals. This has had the effect of blurring ideological boundaries and of having employees accept a certain level of stress and ill-health in order to ensure that their right to work is maintained.

During the past decade or more, government's and to some degree unions' interest in democratization in the workplace has diminished. Instead of attempting to systematically shift the center of control in work organizations to lower levels to ensure a greater need satisfaction and greater goal orientation for employees, ideological and strategic control has moved further into the hands of managers. With the exception of some union- and government-initiated workplace health and safety legislation and programs, the steering force for the future of industrial organization is moving more firmly into the hands of management. This is increasingly supported by government. As planning and control become more distant from employees, the more likely they are to have to accept limited need satisfactions. Apart from positive legislation, the implementation of initiatives for reducing stress and for increasing levels of health at work is likely to be hard fought.

LIMITATIONS OF THE STUDY

These are certain areas that this book has not covered. First, it does not include a number of other variables that interact in the relationship between alienating work, stress, and ill-health. Social support has been shown to play a role in mediating the effects of negative work on stress in some studies. A preliminary analysis of the effects of social support was undertaken in this study, but no significant effects on stress were found and it was therefore omitted from further analysis.

Second, a longitudinal study would have identified changes in work factors that contribute to increased stress and ill-health. However, it was not possible to secure the required cooperation from the firm for an extended period of time.

Third, research (see 101) has shown that variables in the stress and ill-health model can have a variety of effects. For example, self-esteem and personal mastery can be tested for their effects together with ascribed and achieved status variables, prior to occupational membership. That is, the esteem and mastery that people develop prior to their current work experience can have effects on all other variables including occupational position. There was sufficient evidence from previous studies, however, to suggest that work experience helps to establish elements of self-esteem and personal mastery, and these were therefore included as dependent variables together with stress. Their position at an earlier point in the causal model will be tested later.

Fourth, the study did not compare each separate occupational group throughout all stages of the analysis. Small sample sizes for some of the occupational categories would have meant a substantial reduction in the reliability of results, especially through regression analysis.

Finally, the questionnaire gathered substantial data on medical help-seeking and associated behavior. These have not been discussed in this book but will be published later.

FURTHER RESEARCH AND POLICY IMPLICATIONS

The research described in this book has supported the causal link between stress and ill-health and has demonstrated differences in the causes and consequences of stress for different occupational and gender groups. Further research is needed, however, on the nature and direction of the causal links in the stress and ill-health relationship. For instance, the direction and nature of influences from low self-esteem and low mastery need further investigation. In addition, very little research has been conducted on the role played by coping strategies and on the ways in which these mediate both stress and ill-health symptoms. Further study is also needed on the actual nature of the stress response and the processes leading to ill-health. This could be achieved by qualitative, interview-based studies identifying the specific types of work factors that create stress for individuals and for groups of workers. In addition, more qualitative-based studies could investigate more closely how stress manifests and how various factors mediate the experience of stress on ill-health outcomes. This would require more processual types of studies.

The results of this study have certain policy implications. Compensation for ill-health caused by occupational stress will not alleviate the causes of stress, but in order to deal equitably with the consequences of entrenched, inequitable work relations, compensation legislation should be extended. In particular, it needs to ensure that lower level, blue-collar workers are adequately dealt with by compensation tribunals and fairly compensated.

In Australia, earlier federal and state government initiatives that actively encouraged and sought more cooperative working relationships between management and workers should be reactivated in order to start to deal with the fundamental causes of stress. Occupational health should be included in agendas on cooperative work practices as an important part of that process. Scandinavian experiences of greater cooperation attest to stress and health benefits. Bolinder (159), for example, by studying differences in negative attitudes by management toward employees in democratized versus bureaucratized organizations, has shown that these attitudes represent most important sources of stress and ill-health.

Initiatives to encourage the use of innovative management practices such as multi-skilling and quality organization could be combined with other management practices which have the goal of shifting the level of operational and strategic decision-making to lower levels. These could help to alleviate some of the harmful effects of increasing management control under capitalism, especially those that result from more stringent economic conditions.

References

1. Burawoy, M. *The Politics of Production.* Verso, London, 1985.
2. Braverman, H. *Labour and Monopoly Capital: The Degradation of Work in the Twentieth Century.* Monthly Review, New York, 1974.
3. Taylor, F. W. *Scientific Management.* Harper, New York, 1947.
4. Coburn, D. Job alienation and well-being. *Int. J. Health Serv.* 9(1): 41–59, 1979.
5. Garfield, J. Alienated labor, stress and coronary disease, *Int. J. Health Serv.* 10(4): 551–561, 1980.
6. Navarro, V. The labour process and health: A historical materialist interpretation. *Int. J. Health Serv.* 12(1): 5–29, 1982.
7. Schwalbe, M. L., and Staples, C. L. Class position, work experience and health. *Int. J. Health Serv.* 16(4): 583–602, 1986.
8. Schatzkin, A. Health and labour power: A theoretical investigation. *Int. J. Health Serv.* 8(2): 213–234, 1978.
9. Selye, H. A syndrome produced by diverse nocuous agents. *Nature* 138: 32, 1936.
10. Frankenhaeuser, M. Coping with job stress—A psychobiological approach. In *Working Life: A Social Science Contribution to Work Reform,* edited by B. Gardell and G. Johansson, pp. 213–234. Wiley, New York, 1981.
11. Hinkle, L. E. The concept of stress in the biological and social sciences. *Sci. Med. Man* 1: 31–48, 1973.
12. Lumsden, D. P. Towards a systems model of stress: Feedback from an anthropological study of the impact of Ghana's Volta River project. In *Stress and Anxiety,* edited by I. G. Sarson, and C. D. Speilberg, Vol. 2, pp. 191–228. Hemisphere, Washington D.C., 1975.
13. Cox, T., and Mackay, C. A transactional approach to occupational stress. In *Stress, Work Design and Productivity,* edited by E. N. Corlett and J. Richardson. Wiley, Chichester, 1981.
14. Mason, J. W. A historical view of the stress field. Part 1. *J. Hum. Stress* 1(1): 6–12, 1975.
15. Selye, H. *The Stress of Life.* McGraw Hill, New York, 1956.
16. Lazarus, R. S. *Psychological Stress and the Coping Process.* McGraw Hill, New York, 1966.
17. Levi, L. Stress in industry. Causes, effects and prevention. *Occupational Safety and Health,* Ser. No. 51. International Labour Organisation, Geneva, 1984.
18. Otto, R. *Teachers Under Stress: Health Hazards in a Work Role and Modes of Responses.* Hill of Content, Melbourne, 1986.

19. Cannon, W. B. The interrelationships of emotions as suggested by recent physiological researchers. *Am. J. Psychol.* 25: 256–282, 1914.

20. Cannon, W. B. *Bodily Changes in Pain, Hunger, Fear and Rage.* Bramford, Boston, 1929.

21. Selye, H. *Stress in Health and Disease.* Butterworth, Boston, 1976.

22. Pollock, K. On the nature of social stress: Production of a modern mythology. *Soc. Sci. Med.* 26(3): 381–392, 1988.

23. Seggie, J., and Brown, G. M. Profiles of hormone stress response: Recruitment or pathway specificity. In *Brain Neurotransmitters and Hormones,* edited by R. Collu et al., pp. 277–285. Raven Press, New York, 1982.

24. Ferrandez, M. D., Maynar, M., and De la Fuente, M. Effects of long term training program of increasing intensity on the immune function of indoor Olympic cyclists. *Int. J. Sports Med.* 17(8): 592–596, 1996.

25. Huether, G., et al. Psychological stress and neuronal plasticity: An expanded model of the stress reaction as the basis for understanding central nervous system adaptation processes. *Z. Psychosom. Med. Psychoanal.* 42(2): 107–127, 1996.

26. Mason, J. W. A historical view of the stress field. Part 2. *J. Hum. Stress* 1(2): 22–36, 1975.

27. Monet, A., and Lazarus, R. S. (eds.). *Stress and Coping: An Anthology.* Columbia University Press, New York, 1977.

28. Cassell, J. Physical illness in response to stress. In *Social Stress,* edited by S. Levine and N. A. Scotch, pp. 189–209. Aldine, Chicago, 1970.

29. Levi, L. Psychosocial stress and disease: A conceptual model. In *Life Stress and Illness,* edited by E. K. E. Gunderson and R. H. Rahe, pp. 8–33. C. C. Thomas, Springfield, Ill., 1974.

30. Egger, J. From psychological stress to neuropsychoimmunology. *Paediatr. Pathol.* 27(4): 91–96, 1992.

31. Christiansen, N. J., and Jensen, E. W. Effects of psychosocial stress and age on plasma neopinephrine levels: A review. *Psychosom. Med.* 56(1): 77–83, 1994.

32. Melamed, S., and Bruhis, S. The effects of chronic industrial noise exposure on urinary control, fatigue and irritability: A controlled field experiment. *J. Occup. Environ. Med.* 38(3): 252–256, 1996.

33. Wolff, H. G. Disease and patterns of behaviour. In *The Hour of Insight,* edited by R. MacIver, pp. 29–41. Institute for Religious and Social Studies, New York, 1954.

34. Hinkle, L. E., et al. An investigation of the relation between life experience, personality characteristics, and general susceptibility to illness. *Psychosom. Med.* 20: 278–295, 1958.

35. Mason, J. W. The scope of psychoendocrine research. *Psychosom. Med.* 30: 565–575, 1968.

36. Mason, J. W. A review of psychoendocrine research on the pituitary-adrenal corticol system. *Psychosom. Med.* 30: 576–607, 1968.

37. Mason, J. W. A review of psychoendocrine research in the sympathetic-adrenal medullary system. *Psychosom. Med.* 30: 631–653, 1968.

38. Groenink, L., et al. The 5-HTIA receptor is not involved in emotional stress-induced rises in stress hormones. *Pharmacol. Biochem. Behav.* 55(2): 303–308, 1996.

39. Burchfield, S. R. The stress response: A new perspective. *Psychosom. Med.* 41: 661–672, 1979.

40. Vingerhoets, A. A. J. M., and Marcelissen, F. H. G. Stress research: Its present status and issues for future developments. *Soc. Sci. Med.*. 26(3): 279–291, 1988.

41. Henry, J. P., and Stephens, P. M. *Stress, Health and the Social Environment: A Sociobiologic Approach to Medicine.* Springer Verlag, New York, 1977.

42. Weiss, J. M. Psychological factors in stress and disease. *Sci. Am.* 226(6): 104–113, 1972.

43. Baum, A., Gatchel, R. J., and Schheffer, M. A. Emotional, behavioural and physiological effects of chronic stress at Three Mile Island. *J. Consult. Clin. Psychol.* 51(4): 565–572, 1983.

44. Cahill, L., and McGaugh, J. I. Modulation of memory storage. *Curr. Opin. Neurobiol.* 69(2): 237–242, 1996.

45. Weinstein, J., et al. Defensive style and discrepancy between self-reports and physiological indices of stress. *J. Pers. Soc. Psychol.* 10: 406–413, 1968.

46. Bandura, A., et al. Catecholamine secretion as a function of perceived coping self-efficacy. *J. Consult. Clin. Psychol.* 53(3): 406–414, 1985.

47. Bush, I. E. Chemical and biological factors in the activity of adrenocortical steroids. *Pharmacol. Rev.* 14(3): 317–445, 1962.

48. Mason, J. W. "Over-all" hormonal balance as a key to endocrine organisation. *Psychosom. Med.* 30: 791–808, 1968.

49. Mason, J. W. A re-evaluation of the concept of non-specificity in stress theory. *J. Psychiatr. Res.* 8: 323–333, 1971.

50. Vingerhoets, A. J. J. M. *Psychosocial Stress: An Experimental Approach. Life Events, Coping and Psycho-biological Functioning.* Swet and Zeitlinger, Lisse, 1985.

51. Ducharme, J. R., et al. Effects of stress on the hypothalamic–pituitary–testicular function in rats. In *Brain Neurotransmitters and Hormones,* edited by R. Collu, et al., pp. 305–317. Raven Press, New York, 1982.

52. Smeets, H. J., et al. Differential effects of counterregulatory stress hormones on serum albumin concentrations and protein catabolism in healthy volunteers. *Nutrition* 11(5): 423–427, 1995.

53. Donald, R. A., et al. The plasma interleukin-6 and stress hormone responses to acute pyelonephritis. *J. Endocrinol. Invest.* 17(4): 263–268, 1994.

54. Ferraccioli, G., et al. Somatomedin C (insulin-like growth factor 1) levels decrease during acute changes of stress related hormones. Relevance for fibromyalgia. *J. Rheumatol.* 21(7): 1332–1334, 1994.

55. Elenkov, I. J., et al. Modulatory effects of glucocorticoids and catecholamines on human interleukin-12 and interleukin-10 production: Clinical implications. *Proc. Assoc. Am. Physicians* 108(5): 374–381, 1996.

56. McCarthy, R., Horwatt, K., and Konarska, M. Chronic stress and sympathetic-adrenal medullary responsiveness. *Soc. Sci. Med.* 26(3): 333–341, 1988.

57. Brown, G. M., et al. Psychoneuroendocrinology and growth hormone: A review. *Psychoneuroendocrinology* 3: 131–153, 1978.

58. Guillemin, R., et al. B-endorphin and adrenocorticotropin are secreted concomitantly by the pituitary gland. *Science* 197: 1367–1369, 1977.

59. Pedersen, B. K., et al. Exercise-induced immunomodulation—possible roles of neuroendocrine and metabolic factors. *Int. J. Sports Med.* 18(Suppl. 1): S2–57, 1997.

60. Lynch, H. J., Hsun, M., and Wurtman, R. J. Sympathetic neural control of indoleamine metabolism in the rat pineal gland. *Adv. Exp. Med. Biol.* 54: 93–114, 1975.

61. Kopin, I. J., et al. Catecholamines in plasma and response to stress. In *Catecholamines and Stress: Recent Advances,* edited by E. Usdin, R. Kvetnansky, and I. J. Kopin, pp. 197–204. Elsevier, North-Holland, New York, 1980.

62. Pitman, D. L., Natelson, B. H., and Ottenweller, J. E. Classical aversive conditioning of catecholamine and corticosterone responses. *Integr. Physiol. Behav. Sci.* 27(1): 13–22, 1992.

63. Mechanic, D. *Medical Sociology.* Free Press, New York, 1978.

64. Selye, H. Stress without distress in working. In *Society, Stress and Disease: Working Life,* edited by L. Levi, Vol. 4, pp. 263–269. Oxford University Press, New York, 1981.

65. Karasek, R. A. Job demands, job decision latitude and mental strain: Implications for job re-design. *Adminstr. Sci. Q.* 24: 285–308, 1979.

66. Karasek, R. A. Job socialisation and job strain: The implications of two related psychosocial mechanisms for job design. In *Working Life: A Social Science Contribution to Work Reform,* edited by B. Gardell and G. Johansson, pp. 75–94. Wiley, New York, 1981.

67. Mechanic, D. Social structure and personal adaptation: Some neglected dimensions. In *Coping and Adaptation,* edited by G. V. Coelho, D. A. Hamburg, and J. E. Adams, pp. 32–44. Basic Books, New York, 1974.

68. Spillane, R. Stress at work: A review of Australian research. *Int. J. Health Serv.* 14(4): 589–604, 1984.

69. Campbell, F., and Singer, G. *Brain and Behaviour: Psychobiology of Everyday Life.* Pergamon Press, Sydney, 1979.

70. Lazarus, R. S. Physiological stress and coping in adaptation and illness. *Int. J. Psychiatr. Med.* 5: 321–333, 1974.

71. Dohrenwend, B. P., and Dohrenwend, B. S. *Social Status and Psychological Disorder.* Wiley, New York, 1969.

72. Lazarus, R. S., et al. Psychological stress in the workplace. *J. Soc. Behav. Personality* 6(7): 1–13, 1991.

73. Wasserman, G. D. Human behaviour and biology. *Dialectica* 37(3): 169–184, 1983.

74. O'Leary, A. Stress, emotion, and human immune function. *Psychol. Bull.* 108(3): 363–382, 1990.

75. Nichols, T. The sociology of accidents and the social production of industrial injury. In *People and Work,* edited by G. Esland, G. Salaman, and M. Speakman, pp. 217–229. Holmes McDougall, Edinburgh, 1975.

76. Quinlan, M. Psychological and sociological approaches to the study of occupational illness: A critical review. *Austr. N.Z. J. Sociol.* 24(2): 189–207, 1988.

77. Sutherland, V. J., and Cooper, C. L. *Understanding Stress: A Psychological Perspective for Health Professionals.* Chapman and Hall, London, 1990.

78. Dunbar, F. *Psychosomatic Medicine.* Hoeber, New York, 1943.

79. Rosenman, R. H., et al. Coronary heart disease in the Western collaborative Group Study: Final follow up, experience of eight and one half years. *JAMA* 233: 872–877, 1975.

80. Haynes, S. G., Feinleib, M., and Eaker, E. D. Type A behaviour and the ten year incidence of coronary heart disease in the Framington Heart Study. In *Psychosomatic Risk Factors and Coronary Heart Disease: Indication for Specific Preventative Therapy: A Colloquium Held during the International Symposium on Psychophysiological Risk Factors of Cardiovascular Disease,* edited by R. H. Rosenman, pp. 80–92. Hans Huber, Bern, Switzerland, 1983.

81. Matthews, K. A., and Glass, D. C. Type A behaviour, stressful life events, and coronary heart disease. In *Stressful Life Events and Their Contexts,* edited by B. S. Dohrenwend and B. P. Dohrenwend, pp. 167–185. Rutgers University Press, New Brunswick, N.J., 1984.

82. Ham, L. Women's Work, Women's Health: A Study of Occupational Stress in Gender-segregated Sectors of Industry. Ph.D. dissertation, Department of Sociology, La Trobe University, Melbourne, 1992.

83. Caplan, R. D., et al. *Job Demands and Worker Health.* Survey Research Centre, Institute for Social Research, University of Michigan, Ann Arbor, 1980.

84. Lamude, K. G., and Scudder, J. Conflict strategies and Type A scoring managers. *Psychol. Rep.* 71(2): 611–617, 1992.

85. Denolette, J., and de Potter, B. Coping subtypes for men with coronary heart disease: Relationship to well-being, stress and Type A behaviour. *Psychol. Med.* 22(3): 667–684, 1992.

86. Greenglass, E. R., and Burke, R. J. The relationship between stress and coping among Type As. *J. Soc. Behav. Pers.* 6(7): 361–373, 1991.

87. Byrne, D. G., and Reinhart, M. I. Self reported distress, job dissatisfaction and the type A behaviour pattern in a sample of full time employed Australians. *Work Stress* 4(2): 155–166, 1990.

88. Jamal, M. Relationship of job stress and Type A behaviour to employees' job satisfaction, organisational commitment, psychosomatic health problems, and turnover motivation. *Hum. Relations* 43(8): 727–738, 1990.

89. Davilla, D. M., Mariotta, E., and Hicks, R. A. Type A-B behaviour and self reported health problems. *Psychol. Rep.* 67(3): 960–962, 1990.

90. Weiss, M. Components of the Type-A behaviour pattern and their relevance for C.H.D. risk factors: Results from the GDR-Monica project. *Activitas Nervosa Superior* 32(1): 38–40, 1990.

91. Nahavandi, A., Mizzi, P. J., and Malekzadeh, A. R. Executives' Type A personality as a determinant of environmental perception and firm strategy. *J. Soc. Psychol.* 132(1): 59–67, 1992.

92. Barnes, B. L. Stress in transport workers. *Indian J. Clin. Psychol.* 19(1): 14–17, 1992.

93. Dunbar, F. *Mind and Body: Psychosomatic Diagnosis.* Hoeber, New York, 1948.

94. Deary, I. J., et al. Reporting of minor physical symptoms and family incidence of hypertension and heart disease. Relationships with personality and type A behaviour. *Pers. Individual Diff.* 12(7): 747–751, 1991.

95. Freidman, E. S., Clark, D. B., and Gershon, S. Stress, anxiety, and depression: Review of biological, diagnostic, and nosologic issues. *J. Anxiety Disord.* 6(4): 337–363, 1992.

96. Endler, N. S., and Parker, J. D. Stress and anxiety: Conceptual and assessment issues. Special issue: II-IV. Advances in measuring life stress. *Stress Med.* 6(3): 243–248, 1990.

97. Lopez, I., and Juan, J. The meaning of stress, anxiety and collective panic in clinical settings. Ninth World Congress of the International College of Psychosomatic Medicine. *Psychother. Psychosom.* 47(3-4): 168–174, 1987.

98. Pierre, K. D. Enhancing well-being at the workplace: A challenge for EAPs. *Employee Assistance Q.* 1(4): 19–28, 1986.

99. Pearlin, L. I., and Schooler, C. The structure of coping. *J. Health Soc. Behav.* 19: 2–21, 1978.

100. Meier, S. T. Tests of the construct validity of occupational stress measures with college students: Failure to support discriminant validity. *J. Counsel. Psychol.* 38(1): 91–97, 1991.

101. Israel, B. A., et al. The relation of personal resources, participation, influence, and interpersonal relationships and coping strategies to occupational stress, job strain and health: A multivariate analysis. *Work Stress* 3(2): 163–194, 1989.

102. Andries, F., Kompier, M. A. J., and Smulders, P. G. W. Do you think that your health or safety are at risk because of your work? A large European study on psychological and physical work demands *Work Stress* 10(2): 104–118, 1996.

103. Hall, E. M. Gender, work control and stress: A theoretical discussion, and an empirical test. *Int. J. Health Serv.* 19(4): 725–745, 1989.

104. Cooper, C. L., and Marshall, J. Occupational sources of stress: A review of the literature relating to coronary heart disease and mental ill-health. *J. Occup. Psychol.* 49: 11–28, 1976.

105. Perrewe, P. L., and Ganster, D. C. The impact of job demands and behavioural control on experienced job stress. *J. Organizational Behav.* 10(3): 213–229, 1989.

106. Landsbergis, P. A. Occupational stress among health care workers: A test of the job demands-control model. *J. Organizational Behav.* 9(3): 217–239, 1988.

107. Gutterman, N. B., and Jayarante, S. Perceptions of stress, control and professional effectiveness in child welfare district practitioners. *J. Soc. Serv. Res.* 20(1-2): 99–120, 1994.

108. Frankenhaeuser, M., and Johansson, G. Stress at work: Psychological and psychosocial aspects. Special issue: Occupational and life stress and the family. *Int. Rev. Appl. Psychol.* 35(3): 287–299, 1986.

109. Frankenhaeuser, M., and Gardell, B. Underload and overload in working life: Outline of a multidisciplinary approach. *J. Hum. Stress* 2: 35–46, 1976.

110. Broadbent, D. E., and Gath, D. Symptom levels in assembly line workers. In *Machine Pacing and Occupational Stress,* edited by G. Salvendy and M. J. Smith, pp. 243–252. Taylor and Francis, London, 1981.

111. Karmaus, W. Working conditions and health: Social epidemiology, patterns of stress and change. Eighth International Conference on the Social Sciences and Medicine (1983, Stirling, Scotland). *Soc. Sci. Med.* 19(4): 359–372, 1984.

112. Nelkin, D. Workers at risk. *Science* 222(4620): 125, 1983.

113. Warr, P. B. Decision latitude, job demands and employee well-being. *Work Stress* 4(4): 285–294, 1990.
114. Braun, S., and Hollander, R. B. Work and depression among women in the Federal Republic of Germany. *Women Health* 14(2): 3–26, 1988.
115. Theorell, T., et al. A psychosocial and biomedical comparison between men in six contrasting service occupations. *Work Stress* 4(1): 51–63, 1990.
116. Smith, M. J. Occupational stress: An overview of psychosocial factors. In *Machine Pacing and Occupatinal Stress,* edited by G. Salvendy and M. J. Smith, pp. 13–19. Taylor and Francis, London, 1981.
117. Smith, M. J., Hurrell, J., and Murphy, R. K. Stress and health effects in paced and unpaced work. In *Machine Pacing and Occupational Stress,* edited by G. Salvendy and M. J. Smith, pp. 261–268. Taylor and Francis, London, 1981.
118. Wilkes, B., Stammerjohn, L., and Lalich, N. Job demands and worker health in machine paced poultry inspection. *Scand. J. Work Environ. Health Suppl.* 4: 12–19, 1981.
119. Hurrell, J. J., and Colligan, M. J. Machine pacing and shiftwork: Evidence for job stress. Special issue: Job stress: from theory to suggestion. *J. Organizational Behav. Manage.* 8(2): 159–175, 1986.
120. Bulat, V. The human being and paced work on the assembly line. In *Machine Pacing and Occupational Stress,* edited by G. Salvendy and M. J. Smith, pp. 191–196. Taylor and Francis, London, 1981.
121. Franks, I. T., and Sury, R. J. The performance of operators in conveyor-paced work. *Int. J. Production Res.* 5(2): 97–112, 1966.
122. Manenica, I. Comparisons of some physiological indices during paced and unpaced work. *Int. J. Production Res.* 15(3): 161–175, 1977.
123. Cox, T. Repetitive work. In *Current Concerns in Occupational Stress,* edited by C. L. Cooper and R. Payne, pp. 23–41. Wiley, Chichester, 1980.
124. Johansson, G. Subjective well-being and temporal patterns of sympathetic-adrenal medullary activity. *Biol. Psychol.* 4: 157–172, 1976.
125. Hokanson, J. E., et al. Availability of avoidance behaviours in modulating vascular-stress responses. *J. Pers. Soc. Psychol.* 19: 60–68, 1971.
126. Johansson, G. Psychoneuroendocrine correlates of unpaced and paced performance. In *Machine Pacing and Occupational Stress,* edited by G. Salvendy and M. J. Smith, pp. 277–286. Taylor and Francis, London, 1981.
127. Kornhauser, A. *Mental Health of the Industrial Worker.* Wiley, New York, 1965.
128. Caplan, R. D., et al. *Job Demands and Worker Health.* U.S. Department of Health, Education and Welfare, Publication No. (NIOSH) 75-160. U.S. Government Printing Office, Washington, D.C., 1975.
129. Savery, L. K. Stress and the employee. *Leadership Organizational Dev. J.* 7(2): 17–20, 1986.
130. Aronsson, G. Dimensions of control as related to work organization, stress, and health. *Int. J. Health Serv.* 19(3): 459–468, 1989.
131. Houben, G. J. Production control and chronic stress in work organizations. *Int. J. Health Serv.* 21(2): 309–327, 1991.
132. Wright, E. O. *Classes.* New Left, London, 1985.

133. Kohn, M. L., et al. Position in the class structure and psychological functioning in the United States, Japan, and Poland. *Am. J. Sociol.* 95(4): 964–1008, 1990.

134. Otto, R. Patterns of Stress, Symptom Awareness and Medical-Help Seeking among Women and Men in Selected Occupations. Ph.D. dissertation, La Trobe University, Bundoora Campus, Melbourne, 1976.

135. Anderson, W. J. R., Cooper, C. L., and Willmott, M. Sources of stress in the National Health Service: A comparison of seven occupational groups. A Research Note. *Work Stress* 10(1): 88–95, 1996.

136. Barney, J. A., et al. An analysis of social class and certain attitudinal and personality variables: A comparison. *College Student J.* 21(1): 13–18, 1987.

137. Frankenhaeuser, M., and Rissler, A. Effects of punishment on catecholamine release and efficiency of performance. *Psychopharmacologia* 17: 378–390, 1970.

138. French, J. R. P., and Caplan, R. D. Organisation stress and individual strain. In *Failure of Success*, edited by A. J. Murrow, pp. 30–66. American Management Association, New York, 1973.

139. Gardell, B. Autonomy and participation at work. In *Society, Stress and Disease: Working Life*, edited by L. Levi, pp. 279–289. Oxford University Press, Oxford, 1981.

140. French, J. R. P., and Caplan, R. D. Psychosocial factors in coronary heart disease. *Ind. Med. Surg.* 39: 383–397, 1970.

141. Emery, F. E., and Thorsrud, E., in co-operation with Trist, E. *Form and Content in Industrial Democracy: Some Experiences from Norway and other European Countries.* Tavistock, London, 1969.

142. Jackson, S. E. Participation and decision making as a strategy for reducing job-related strain. *J. Appl. Psychol.* 68: 3–19, 1983.

143. Kommer, M. M. A Dutch prison officer's work: Balancing between prison policy, organisational structure and professional autonomy. *Neth. J. Soc. Sci.* 29(2): 130–146, 1993.

144. O'Brien, G. E., Dowling, P., and Kabanoff, B. *Work, Health and Leisure.* Working Paper No. 28. National Institute of Labour Studies, Adelaide, 1978.

145. Van der Auwera, F. Working in a network: Convergences and divergences in the labor market [wer in een netwerk: convergenties en divergenties in de arbeidsmarkt]. *Tijdschrift voor sociologie* 6(1-2): 103–123, 1985.

146. Lowe, G. S., and Northcott, H. C. The impact of working conditions, social roles, and personal characteristics on gender differences in distress. *Work Occup.* 15(1): 55–77, 1988.

147. Kahn, R. L., et al. *Organisational Stress: Studies in Role Conflict and Ambiguity.* Wiley, New York, 1964.

148. Martin, T. N. Role stress and inability to leave as predictors of mental health. *Hum. Relations* 37(11): 969–983, 1984.

149. Travers, C. J., and Cooper, C. L. Mental health, job satisfaction and occupational stress among U.K. teachers. *Work Stress* 7(3): 203–219, 1993.

150. Hammer, W. C., and Tosi, H. L. Relationships of role conflict role ambiguity to job involvement measures. *J. Appl. Psychol.* 59: 497–499, 1974.

151. Kahn, R. L. Conflict, ambiguity and overload: Three elements in job stress. In *Occupational Stress,* edited by A. MacLean, pp. 47–61. C. C. Thomas, Springfield, Ill., 1974.

152. Sutherland, V., and Davidson, M. J. Using a stress audit: The construction site manager's experience in the U.K. *Work Stress* 7(3): 273–286, 1993.

153. Sutherland, V. J., and Cooper, C. L. Identifying distress among general practitioners: Predictors of psychological ill-health and job dissatisfaction. *Soc. Sci. Med.* 37(5): 575–581, 1993.

154. Leigh, J. H., Lucas, G. H., and Woodman, R. W. Effects of perceived organisational factors on role stress–job attitude relationships. *J. Manage.* 14(1): 41–58, 1988.

155. Guppy, A., and Gutteridge, T. Job satisfaction and occupational stress in U.K. general hospital nursing staff. *Work Stress* 5(4): 315–323, 1991.

156. Cooper, C. L., and Marshall, J. Success of managerial and white collar stress. In *Stress at Work,* edited by C. L. Cooper and R. Payne, pp. 81–105. Wiley, Chichester, 1978.

156. van Dijkhuizen, N. Towards organisational coping with stress. In *Coping with Stress at Work: Case Studies from Industry,* edited by J. Marshall and C. L. Cooper, pp. 203–220. Gower, London, 1978.

158. Buck, V. E. *Working under Pressure.* Staples Press, London, 1972.

159. Bolinder, E. Stress and disease: From the viewpoint of a confederation of trade unions. In *Society, Stress and Disease: Working Life,* edited by L. Levi, Vol. 4, pp. 28–38. Oxford University Press, Oxford, 1981.

160. Kirkcaldy, B. D., and Cooper, C. L. Cross cultural differences in occupational stress among British and German managers. *Work Stress* 6(2): 177–190, 1992.

161. Child, J. Managerial strategies, new technology and the labour process. In *On Work: Historical, Comparative and Theoretical Approaches,* edited by R. E. Pahl, pp. 229–257. Blackwell, Oxford, 1988.

162. Littler, C. R., and Salaman, G. Bravermania and beyond: Recent theories of the labour process. *Sociology* 16(2): 251–269, 1982.

163. Masreliez, N. Psychosocial stressors in the work environment and in non-work settings: Management's viewpoints. In *Society, Stress and Disease: Working Life,* edited by L. Levi, Vol. 4, pp. 24–27. Oxford University Press, Oxford, 1981.

164. McLean, A. *Work Stress.* Addison Wesley, Reading, Mass., 1979.

165. Siegrist, J., and Klein, D. Occupational stress and cardiovascular reactivity in blue collar workers. *Work Stress* 4(4): 295–304, 1990.

166. Endresen, I. M., et al. Stress at work and psychological and immunological parameters in a group of Norwegian female bank employees. *Work Stress* 5(3): 217–227, 1991.

167. Richardsen, A. M., and Burke, R. J. Occupational stress and job satisfaction among Canadian physicians. *Work Stress* 5(4): 301–313, 1991.

168. Miles-Tapping, C. Caring in profit: Alienation and work stress in nursing assistants in Canada. *Work Stress* 6(1): 3–12, 1992.

169. Petitone, A. Self, social environment and stress. In *Psychological Stress,* edited by M. H. Appley and R. Trumbull, pp. 182–199. Appleton-Century-Crofts, New York, 1967.

170. Breslow, I., and Buell, P. Mortality from coronary heart disease and physical activity at work in California. *J. Chronic Dis.* 11: 615–626, 1960.
171. Kirmeyer, S. L., and Dougherty, T. W. Work load, tension, and coping: Moderating effects of supervisor support. *Personnel Psychol.* 41(1): 125–139, 1988.
172. Rissler, A. Stress reactions at work and after work during a period of quantitative overload. *Ergonomics.* 20: 13–16, 1977.
173. Bruner, B. M., and Cooper, C. L. Corporate financial performance and occupational stress. *Work Stress* 5(4): 267–287, 1991.
174. Ratsoy, E. W., Sarros, J. C., and Aidoo, T. N. Organisational stress and coping: A model and empirical check. *Alberta J. Educ. Res.* 32(4): 270–285, 1986.
175. Del'Erba, G., Pancheri, P., and Intreccialagli, B. Hormonal assessment and workload correlates in air traffic controllers at the end of night shift: The stress perspective. *New Trends Exp. Clin. Psychiatry* 4(4): 197–212, 1988.
176. Staples, C. L., Schwalbe, M. L., and Gecas, V. Social class, occupational conditions and efficacy based self-esteem. *Sociol. Perspect.*. 27(1): 85–109, 1984.
177. Schwalbe, M. L. Sources of self-esteem in work: What's important for whom? *Work Occup.* 15(1): 24–35, 1988.
178. Pearlin, L. I., et al. The stress process. *J. Health Soc. Behav.* 22: 337–356, 1981.
179. Margolis, B. L., and Kroes, W. H. *Occupational stress and strain. Occup. Ment. Health* 2(4): 4–6, 1973.
180. Rees, D. W., and Cooper, C. L. The Occupational Stress Indicator locus of control scale: Should this be regarded as a state rather than a trait measure? *Work Stress* 6(1): 45–48, 1992.
181. Daniels, K., and Guppy, A. Control, information-seeking preferences, occupational stressors and psychological well-being. *Work Stress* 6(4): 347–353, 1992.
182. Osipow, S. H., Doty, R. E., and Spokane, A. R. Occupational stress, strain, and coping across the life span. *J. Vocational Behav.* 27(1): 98–108, 1985.
183. Barnes-Farrell, J. L., and Piotrowski, M. J. Discrepancies between chronological age and personal age as a reflection of unrelieved worker stress. *Worker Stress* 5(3): 177–187, 1991.
184. Hinkle, L. E., et al. Coronary heart disease. *Arch. Environ. Health* 13: 312–321, 1966.
185. Wan, T. Status, stress and morbidity: A sociological investigation of selected categories of work-limiting chronic condition. *J. Chronic Dis.* 24: 453–468, 1971.
186. Kleitzman, S., et al. Work stress, nonwork stress, and health. *J. Behav. Med.* 13(3): 221–243, 1990.
187. Bolger, N., et al. The contagion of stress across multiple roles. *J. Marriage Fam.* 51(1): 175–183, 1989.
188. Hall, E. M. Double exposure: The combined impact of home and work environments on psychosomatic strain in Sweden women and men. *Int. J. Health Serv.* 22(2): 239–260, 1992.
189. Burke, R. J. Work-family stress, conflict, coping and burnout in police officers. *Stress Med.* 9(3): 171–180, 1993.
190. Jones, F., and Fletcher, B. C. Taking work home: A study of daily fluctuations in work stressors, effects on moods and impacts on marital partners. *J. Occup. Organizational Psychol.* 69(1): 89–106, 1996.

191. Leiter, M. P., and Durup, M. J. Work, home, and in-between: A longitudinal study of spillover. *J. Appl. Behav. Sci.* 32(1): 29–47, 1996.
192. Avison, W. R. Roles and resources: The effects of family structure and employment on women's psychosocial resources and psychological distress. *Res. Community Mental Health* 8: 233–256, 1995.
193. Gill, G. J., and Hibbins, R. Wives' encounters: Family work stress and leisure in two-job families. *Int. J. Sociol. Fam.* 26(2): 43–54, 1996.
194. Shipley, P., and Coates, M. A community study of dual-role stress and coping in mothers. *Work Stress* 6(1): 49–63, 1992.
195. Shostack, A. B. Blue collar worker alienation. In *Job Stress and Blue Collar Work,* edited by C. L. Cooper and M. J. Smith, pp. 7–18. Wiley, Chichester, 1985.
196. Shirom, A., and Kirmeyer, S. The effects of unions on blue collar role stresses and somatic strain. *J. Organizational Behav.* 9(1): 29–42, 1988.
197. Marmot, M., and Theorell, T. Social class and cardiovascular disease: The contribution of work. *Int. J. Health Serv.* 18(4): 659–674, 1988.
198. Schabroeck, J., Cotton, J. L., and Jennings, K. R. Antecedents and consequences of role stress: A covariance structure analysis. *J. Organizational Behav.* 10(1): 35–58, 1989.
199. Gumenyuk, V. A., et al. Effects of a combination of measures to abolish nervous and emotional strain on cardiorespiratory parameters of factory workers. *Hum. Physiol.* 14(4): 290–297, 1988.
200. Ekberg, K., et al. Psychological stress and muscle activity during data entry at visual display units. *Work Stress* 9(4): 475–490, 1995.
201. Arora, S. A contemporary study of VDU-users and VDU-non users on stress, alienation and physical health. *Abhigyan,* 1994, pp. 39–41.
202. Balshem, M. The clerical workers' boss: An agent of job stress. *Hum. Organization* 47(4): 361–367, 1988.
203. Turnage, J. J., and Speilberger, C. D. Job stress in managers, professionals and clerical workers. *Work Stress* 5(3): 165–176, 1991.
204. Kazin, E. A., et al. Use of automated systems to assess functional states: IV. Role of functional asymmetry parameters in assessment of psychophysiological adaption of administrative workers. *Hum. Physiol.* 18(3): 173–179, 1992.
205. Westman, M. Moderating effect of decision latitude on stress-strain relationship: Does organisational level matter? *J. Organisational Behav.* 13(7): 713–722, 1992.
206. Bharati, T., Nagarathnamma, B., and Reddy, S. V. Effect of occupational stress on job satisfaction. Second Asian Conference on Behavioural Toxicology and Environmental Management (1992, Kanpur, India). *J. Indian Acad. Appl. Psychol.* 17(1-2): 81–85, 1991.
207. Marshall, J., and Cooper, C. L. The causes of managerial job stress: A research note on methods and initial findings. In *Stress, Work Design and Productivity,* edited by E. N. Corlett, and J. Richardson, pp. 115–128. Wiley, Chichester, 1981.
208. Moss, L. *Management Stress.* Addison-Wesley, Reading, Mass., 1981.
209. Minzberg, H. *The Nature of Managerial Work.* Harper and Row, New York, 1973.
210. Das, G. S. Organisational determinants of anxiety based managerial stress. *Vikalpa* 7(3): 217–221, 1982.

211. Blauner, R. *Alienation and Freedom: The Factory Worker and His Industry*. University of Chicago Press, Chicago, 1964.
212. Burrell, G., and Morgan, G. *Sociological Paradigms and Organisational Analysis*, Gibson, Burrell, and Gareth Morgan, England, 1979.
213. Brown, J. A. C. *The Social Psychology of Industry*. Penguin, Harmondsworth, 1970.
214. Arnold, J., Gleeson, F., and Peterson, C. *Moving into Management*. Swinburne Press, Hawthorn, 1991.
215. Hall, A. The corporate construction of occupational health and safety: A labour process, *Canadian Journal of Sociology* 18(1): 1–20, 1993.
216. Elling, R. *The Struggle for Workers' Health: A Study of Six Industrialized Countries*. Baywood, Amityville, N.Y., 1986.
217. Mayhew, C., and Peterson, C. L. *Occupational Health and Safety in Australia: Industry, Small Business and the Public Sector*, Allen & Unwin, St Leonards, 1998, in press.
218. Marx, K. *Capital*. Progress Publishers, Moscow, 1967.
219. Weber, M. *The Theory of Social and Economic Organisation*, translated by A. M. Henderson and T. Parsons. Free Press, New York, 1964.
220. Wright, E. O. *Class Counts*. Cambridge University Press, Cambridge, 1997.
221. Thomas, W. I. The defintion of the situation. In *Sociological Theory*, edited by L. A. Coser and B. Rosenberg, pp. 245–247. Macmillan, New York, 1971.
222. Cooley, C. H. *Human Nature and the Social Order*. Schocken Books, New York, 1964.
223. Volkart, E. H. (ed.). *Social Behaviour and Personality: Contributions of W. I. Thomas to Theory and Social Research*. Social Research Committee, New York, 1951.
224. Giddens, A. *Sociology*. Polity Press, Cambridge, 1989.
225. Wolff, H. G. *Stress and Disease*, revised and edited by S. Wolf and H. Goodall. C. C. Thomas, Springfield, Ill., 1968.
226. Davis, A., and George, J. G. *States of Health and Illness in Australia*. Harper and Row, New York, 1993.
227. Peterson, C. L. Contributions of sociological approaches to the concept of stress: A reference to occupational stress. *Annu. Rev. Health Soc. Sci.*, 1994, pp. 79–91.
228. Frankenhaeuser, M., and Johansson, G. Task demand as reflected in catecholamine excretion and heart rate. *J. Hum. Stress* 2: 15–23, 1976.
229. Frankenhaeuser, M., and Lundberg, U. The influence of cognitive set on performance and arousal under different noise loads. *Motivation Emotion* 1: 139–149, 1977.
230. Cox, T. *Stress*. Macmillan, London, 1978.
231. McGrath, J. (ed.). *Social and Psychological Factors in Stress*. Holt, Rinehart and Winston, New York, 1970.
232. Cox, T. Stress, coping and problem solving. *Work Stress* 1(1): 5–14, 1987.
233. Eyer, J. Prosperity as a cause of death. *Int. J. Health Serv.* 7(1): 125–150, 1977.
234. Lupton, G. M., and Najman, J. *Sociology of Health and Illness: Australian Readings*. MacMillan Educational, South Melbourne, 1995.
235. Mathers, C. *Health Differentials Among Adult Australians Aged 25–64 Years*. AGPS, Canberra, 1994.
236. Doyal, L., with Pennell, I. *The Political Economy of Health*. Pluto Press, London, 1979.

237. Marx, K. *Capital: A Critique of Political Economy,* translated by B. Fowkes, Vol. 1. Penguin, Harmondsworth, 1976.

238. Thompson, P. *The Nature of Work: An Introduction to Debates on the Labour Process.* Macmillan, London, 1983.

239. Burawoy, M. Thirty years of making out. In *On Work: Historical, Comparative and Theoretical Approaches,* edited by R. E. Pahl, pp. 190–209. Blackwell, New York, 1988.

240. Carter, B. A growing divide: Marxist class analysis and the labour process. *Capital Class* 55: 33–72, 1995.

241. Willis, E. (ed.). *Technology and the Labour Process: Australasian Case Studies.* Allen and Unwin, Sydney, 1988.

242. Littler, C. R., and Salaman, G. *Class at Work: The Design, Allocation and Control of Jobs.* Batesford Academic and Educational, London, 1984.

243. Burawoy, M. Towards a Marxist theory of the labour process: Braverman and beyond. *Polit. Soc.* 8(3-4): 247–312, 1978.

244. Hill, S. *Competition and Control at Work.* The New Industrial Sociology, Cambridge, Mass., 1981.

245. Navarro, V. The politics of health care reform in the United States, 1992–94: A historical review. *Int. J. Health Serv.* 25(2): 185–201, 1995.

246. Greenlund, K. J., and Elling, R. H. Capital sectors and workers' health and safety in the United States. *Int. J. Health Serv.* 25(1): 101–116, 1995.

247. Tanner, J., Davies, S., and O'Grady, B. Immanence changes everything: A critical comment on the labour process and class consciousness. *Sociology* 26(3): 439–453, 1992.

248. Hodson, R. The active workers: Compliance and autonomy in the workplace. *J. Contemp. Ethnog.* 20(1): 47–78, 1991.

249. Sharma, A. M. Workers' participation, self management and workers' control. *Indian J. of Soc. Work* 51(2): 279–290, 1990.

250. Sewell, G., and Wilkinson, B. Someone to watch over me: Surveillance, discipline and the just-in-time labour process. *Sociology* 26(2): 271–289, 1992.

251. Russell, B. The subtle labour process and the great skill debate: Evidence from a potash mine-mill operation. *Can. J. Sociol.* 20(3): 359–385, 1995.

252. Boswell, T. Control of work and worker control: A conflict theory of the governance of transaction costs in work relations. *Res. Soc. Stratification Mobil.* 7: 135–162, 1988.

253. Story, J. The means of management control. *Sociology* 19(2): 193–211, 1985.

254. Kivinens, M. The new middle classes and the labour process. *Acta Sociologica* 32(1): 53–73, 1989.

255. Edwards, R. *Contested Terrain: The Transformation of the Workplace in the Twentieth Century.* Basic Books, New York, 1979.

256. Berlinguer, G., Falzi, G., and Figà-Talamanca, I. Ethical problems in the relationship between health and work. *Int. J. Health Serv.* 26(1): 147–171, 1996.

257. Rosenberg, M. *Society and the Adolescent Self Image.* Princeton University Press, Princeton, N.J., 1965.

258. Baldamus, W. *Efficiency and Effort: An Analysis of Industrial Administration.* Tavistock, London, 1961.

259. Gramsci, A. *Prison Notebooks.* International Publishers, New York, 1971.
260. Willis, E. The industrial relations of occupational health and safety: A labour process approach. *Labor Ind.* 2(2): 317–333, 1989.
261. Stilwell, F. *The Accord . . . and Beyond: The Political Economy of the Labor Government.* Pluto Press, Sydney, 1986.
262. Lane, C. Capitalism and culture? A comparative analysis of the position in the labour process and labour market of lower white-collar workers in the financial services sector of Britain and the federal republic of Germany. *Work Employment Soc.* 1(1): 57–83, 1987.
263. Pearse, W., and Refshauge, C. Worker health and safety in Australia: An overview. *Int. J. Health Serv.* 17(4): 635–650, 1987.
264. Quinlan, A., and Bohle, P. *Managing Occupational Health and Safety in Australia: A Multidisciplinary Approach.* MacMillan Educational, South Melbourne, 1991.
265. Biggins, D. R. Occupational health in Australia. *Annu. Rev. Health Soc. Sci.* 5: 115–136, 1995.
266. Quinlan, M. (ed.). *Work and Health: The Origins, Management and Regulation of Occupational Illness.* MacMillan Educational, South Melbourne, 1993.
267. Matthews, J. *Health and Safety at Work: Australian Trade Union Safety Representatives Handbook.* Pluto Press, Sydney, 1994.
268. Willis, E. *Medical Dominance: The Division of Labour in Australian Health Care.* Allen and Unwin, Sydney, 1983.
269. Illich, I. *Limits to Medicine, Medical Nemesis: The Expropriation of Health.* Penguin, Harmondsworth, 1978.
270. Dawson, R. Aspects of workers' compensation stress cases in the administrative appeals tribunal. *J. Occup. Health Safety Aust. N.Z.* 1(1): 55–59, 1985.
271. Figlio, K. How does illness mediate social relations: Workmen's compensation and medico-legal practices 1890–1940? In *The Problem of Medical Knowledge: Examining the Social Construction of Medicine,* edited by P. Wright and A. Treacher, pp. 174–244. Edinburgh University Press, Edinburgh, 1982.
272. Carson, K., and Henenberg, C. The political economy of legislative change: Making sense of Victoria's new occupational health and safety legislation. *Law in Context* 6(2): 1–19, 1988.
273. Deves, L. A. Policy and action in occupational health: A question of ideology. *J. Occup. Health Safety Aust. N.Z.* 5(2): 109–113, 1989.
274. Deves, L. A., and Spillane, R. M. Occupational health, stress, and work organization in Australia. *Int. J. Health Serv.* 19(2): 351–363, 1989.
275. Lam, T. H., et al. Mental health and work stress in office workers in Hong Kong. *Occup. Med.* 27(3): 199–205, 1985.
276. Gibson, F., McGrath, A., and Reid, N. Occupational stress in social work. *Br. J. Soc. Work* 19: 1–16, 1986.
277. Cairns, E., Wilson, R., and McLelland, R. The Validity of the General Health Questionnaire in Community Settings in Northern Ireland. Unpublished report [cited in reference 276].
278. Graetz, B. Health Consequences of Employment and Unemployment. Paper presented at the Australia Longitudinal Survey: Social and Economic Policy Research Conference, December 10–11, 1990.

279. Banks, M. H., et al. The use of the general health questionnaire as an indicator of mental health in occupational studies. *J. Occup. Psychol.* 53: 187–194, 1980.

280. Alwin, D. F. The use of factor analysis in the construction of linear composites in social research. *Sociol. Methods Res.* November 1973, pp. 191–214.

281. Coopersmith, S. *Self-esteem Inventories.* Consulting Psychologists Press, Calif., 1991.

282. Wylie, R. *The Self Concept.* University of Nebraska Press, Lincoln, 1974.

283. Silber, E., and Tippett, J. S. Critical assessment and measurement validation. *Psychol. Rep.* 16: 1017-1071, 1965.

284. Dorn, H. Methods of measuring incidence and prevalence of disease. *Am. J. Public Health* 41: 271–278, 1951.

285. Broadhead, P. Surveying health—A sociological perspective. *Aust. N.Z. J. Sociol.* 23(1): 104–113, 1987.

286. Anderson, R. A Behavioural Model of Families' Use of Health Services. Centre for Health and Administrative Studies, Research Series 25, University of Chicago, 1968.

287. Braunwald, E. Chest discomfort and palpitation. In *Harrison's Principles of Internal Medicine,* edited by E. Braunwald et al., pp. 17–23. McGraw-Hill, New York, 1987.

288. Cassem, E. H. Approaches to the patient with mental and emotional complaints. In *Harrison's Principles of Internal Medicine,* edited by E. Braunwald et al., pp. 60–64. McGraw-Hill, New York, 1987.

289. Mankin, H. J., and Adams, R. D. Pain the back and neck. In *Harrison's Principles of Internal Medicine,* edited by E. Braunwald et al., pp. 34–42. McGraw-Hill, New York, 1987.

290. Isselbacher, K. J., and May, R. J. Approaches to the patient with gastrointestinal disease. In *Harrison's Principles of Internal Medicine,* edited by E. Braunwald et al., pp. 1223–1225. McGraw-Hill, New York, 1987.

291. Austen, F. K. Diseases of the immediate type hypersensitivity. In *Harrison's Principles of Internal Medicine,* edited by E. Braunwald et al., pp. 1407–1414. McGraw-Hill, New York, 1987.

292. Goodman, L. S., and Gilman, A. *The Pharmacological Basis of Therapeutics.* Macmillan, New York, 1975.

293. Degoratis, L. *SCL-90 Administration, Scoring and Procedures Manual.* Johns Hopkins Press, Baltimore, Md., 1977.

294. Goldberg, D. P. *Manual of the General Health Questionnaire.* NFER-Nelson, Windsor, 1978.

295. Tennant, C. The general health questionnaire: A valid index of psychological impairment in Australian populations. *Med. J. Aust.* 2: 392–394, 1977.

296. Graetz, B. Multidimensional properties of general health questionnaire. *Soc. Psychiatry Psychiatr. Epidemiol.* 26: 132–138, 1991.

297. Benjamin, S., Decalma, P., and Harran, D. Community screenings for mental illness: A validity study of the general health questionnaire. *Br. J. Psychiatry.* 140: 174–180, 1982.

298. Goldberg, D. P. *The Detection of Psychiatric Illness by Questionnaire.* Maudsley Monographs, No. 21. Oxford University Press, Oxford, 1972.

299. Goldberg, D. P., et al. A comparison of two psychiatric screening tests. *Br. J. Psychiatry* 129: 61–67, 1976.

300. Siegert, R. J., et al. An examination of reported factor structures of the general health questionnaire and the identification of a stable replicable structure. *Aust. J. Psychol.* 39(1): 89–100, 1987.

301. Worsley, A., and Gribbin, C. C. A factor analytic study of the twelve item general health questionnaire. *Aust. N.Z. J. Psychiatry.* 11: 269–272, 1977.

302. Game, A., and Pringle, R. *Gender at Work.* Allen and Unwin, Sydney, 1983.

303. Graetz, B., and McAllister, I. *Dimensions of Australian Society.* MacMillan, South Melbourne, 1988.

304. Dometrius, N. C. *Social Statistics Using SPSS.* Harper Collins, New York, 1992.

305. Pedhauser, E. J. *Multiple Regression in Behavioural Research.* Holt, Reinhart and Winston, New York, 1982.

306. Baxter, J., Gibson, D., and Lynch-Blosse, M. *Double Take.* Australian Government Printing Service, Canberra, 1990.

307. House, J. S. Occupational stress and coronary heart disease: A review and theoretical integration. *J. Health Soc. Behav.* 15: 12–27, 1974.

308. Waitzkin, H. B., and Waterman, B. *The Exploitation of Illness in Capitalist Society.* Bobbs-Merrill, Indianapolis, 1974.

309. Thorell, T., et al. Endocrine markers during a job intervention. *Work Stress* 9(1): 67–76, 1995.

310. Barton, J., et al. The standard shiftwork index: A battery of questionnaires for assessing shiftwork-related problems. *Work Stress* 9(1): 4–30, 1995.

311. Guppy, A., and Rick, J. The influence of gender and grade on perceived work stress and job satisfaction in white collar employees. *Work Stress* 10(2): 154–164, 1996.

312. Pattison, H., and Gross, H. Pregnancy, work and women's well-being: A review. *Work Stress* 10(1): 72–87, 1996.

313. Gross, G. R., et al. Gender differences in occupational stress among correctional officers. *Am. J. Criminal Justice* 19(2): 219–234, 1994.

314. Greenglass, E. R. Gender, work, stress, and coping: Theoretical implications. *J. Soc. Behav. Pers.* 10(6): 121–134, 1995.

315. Dekker, I., and Barling, J. Workforce size and workforce related role stress. *Work Stress* 9(1): 45–54, 1995.

316. Menaghan, E. G., and Merves, E. S. Coping with occupational problems: The limits of individual efforts. *J. Health Soc. Behav.* 25: 406–423, 1984.

317. Gutterman, N. B., and Jayaratne, S. "Responsibility at risk": Perceptions of stress, control and professional effectiveness in child welfare direct practitioners. *J. Soc. Serv. Res.* 20(1-2): 99–120, 1994.

318. Peterson, C. L. Work factors and stress: A critical review. *Int. J. Health Serv.* 24(3): 495–519, 1994.

319. Graham, N. M. H. Psychological stress as a public health problem: How much do we know? *Community Health Stud.* XII(2): 151–160, 1988.

320. Corlett, E. N. Pain, posture and performance. In *Work Design and Productivity,* edited by E. N. Corlett and J. Richardson, pp. 27–42. Wiley, Chichester, 1981.

321. Australian Bureau of Statistics. *1989-90, National Health Survey Preliminary Estimates.* Catalogue No. 4361.0. Government Printer, Canberra, 1991.

322. Australian Bureau of Statistics. *Australian Health Survey, 1983, Preliminary.* Catalogue No. 4348.0. Commonwealth Government Printer, Canberra, 1984.

323. Powles, J., and Salzberg, M. Work class or life-style? Explaining inequities in health. In *Sociology of Health and Illness: Australian Readings,* edited by G. M. Lupton and J. M. Najman, pp. 135–168. MacMillan, South Melbourne, 1989.

324. Otto, R. Occupational Stress Amongst Factory Workers: A Study of a Restricted Sample of Men and Women in Two Factories. La Trobe University Working Papers on Sociology, No. 58. La Trobe University, Department of Sociology, School of Social Science, Bundoora Campus, Melbourne, 1980.

325. Lupton, G. M., and Najman, J. M. Sociology and health care. In *Sociology of Health and Illness: Australian Readings,* edited by G. M. Lupton and J. M. Najman, pp. 1–19. MacMillan, South Melbourne, 1989.

326. Broom, D. Masculine medicine, feminine illness: Gender and health. In *Sociology of Health and Illness: Australiana Readings,* edited by G. M. Lupton and J. M. Najman, pp. 121–134. MacMillan, South Melbourne, 1989.

327. Hraba, J., et al. Gender and well-being in the Czech republic. *Sex Roles* 34(7-8): 517–533, 1996.

328. Salvendy, G., and Pilitsis, J. Psychophysiological aspects of paced and unpaced performance as influenced by age. *Ergonomics* 14: 703–711, 1971.

329. Akabas, S. H. Women, work and mental health: Room for improvement. *J. Primary Prevention* 9(1-2): 130–140, 1988.

330. Winfield, H. R., et al. The general health questionnaire: Reliability and validity for Australian youth. *Aust. N.Z. J. Psychiatry* 23: 53–58, 1989.

331. Graetz, B. Health consequences of employment and unemployment. *Soc. Sci. Med.* 36(6): 715–724, 1993.

332. McPherson, A., and Hall, W. Psychiatric impairment, physical health and work values among unemployed and apprenticed young men. *Aust. N.Z. J. Psychiatry* 17: 335–340, 1983.

333. Olk, M. E., and Friedland, M. L. Trainees experience of role conflict and role ambiguity in supervisory relationships. *J. Counsel. Psychol.* 39(3): 389–397, 1992.

334. Girodo, M. Personality, job stress, and mental health in undercover agents: A structural equation analysis. *J. Soc. Behav. Pers.* 6(7): 375–390, 1991.

335. Warr, P. The measurement of well-being and other aspects of mental health. *J. Occup. Psychol.* 63(3): 193–201, 1990.

336. Srivastava, A. K. Moderating effect of self-actualisation of the relationship of role stress with job anxiety. *Psychol. Stud.* 34(2): 106–109, 1989.

337. Martin, R., and Wall, T. D. Double machine operations and psychological strain. *Work Stress* 3(4): 323–326, 1989.

338. Sharma, S., and Sharma, D. Organisational climate, job satisfaction and job anxiety. *Psychol. Stud.* 34(1): 21–27, 1989.

339. Schlenker, J. A., and Gutek, B. A. Effects of role loss on work-related attitudes. *J. Appl. Psychol.* 72(2): 287–293, 1987.

340. Repetti, R. L. Individual and common components of the social environment at work and psychological well-being. *J. Pers. Soc. Psychol.* 52(4): 710–720, 1987.

341. Launay, G., and Fielding, P. J. Stress among prison officers: Some empirical evidence based on self report. *The Howard J.* 28(2): 138–148, 1989.
342. Parkes, K. R. Mental health in the oil industry: A comparative study of onshore and offshore employees. *Psychol. Med.* 22: 997–1009, 1992.
343. Spetlin, E., et al. The relationship between coping strategies and GHQ-scores in nurses. *Ergonomics* 36(1): 227–232, 1993.
344. Viinamaki, H., et al. Unemployment and mental wellbeing: A factory closure study in Finland. *Acta Psychiatry Scand.* 88: 429–433, 1993.

Index

271